THE CURVATURE OF SPACETIME

The Curvature of Spacetime

NEWTON, EINSTEIN, AND GRAVITATION

Harald Fritzsch

Translated by Karin Heusch

Columbia University Press New York

Columbia University Press
Publishers Since 1893
New York   Chichester, West Sussex

Originally published as *Die Verbogene Raum-Zeit*,
copyright © 1996 Piper Verlag GmbH. München;
English-language translation copyright © 2002
Columbia University Press
All rights reserved

Library of Congress Cataloging-in-Publication Data
Fritzsch, Harald, 1943–
    [Verbogene Raum-Zeit. English]
    The curvature of spacetime : Newton, Einstein, and
    gravitation / Harald Fritzsch ;
    translated by Karin Heusch.
    p. ; cm.
    Includes bibliographical references and index.
    ISBN 0-231-11820-1 (acid free paper)
    1. General relativity (Physics). 2. Gravitation.
    3. Space and time. I. Title

QC 173.6 .F7513 2002
530.11—dc21   2002018809

Printed in the United States of America
Designed by Audrey Smith

c 10 9 8 7 6 5 4 3 2 1

The theory presented here is the most ambitious generalization imaginable of what we call the "theory of relativity" in today's common usage. To tell that one apart from what we show here, I will henceforth call it the "special theory of relativity." I will assume it to be known to the reader. To put it into a more general framework, I was much helped by the form it was given by the mathematician [Hermann] Minkowski, who was the first to point out the formal equivalence of the space and time coordinates, and who used this equivalence in constructing our theory.

—Albert Einstein (at the beginning of his article on the foundations of the general theory of relativity, in the *Annals of Physics* 51, 4th series, 1916)

Contents

Preface

In the early 1920s, a journalist asked the famous British astronomer and astrophysicist Arthur Eddington, well known for his witticisms, if it were true that only three people in the world understand the general theory of relativity. When Eddington hesitated to give an answer, the journalist interpreted this as extreme modesty and repeated the question. Thereupon Eddington said: "I'm just asking myself who the third person might be."

Undoubtedly, Eddington understated the case since, at that time, several dozen physicists and mathematicians were working intensely on Einstein's theory. Today, the basics of the general theory of relativity are known—at least in its rough outlines—to most physicists. But one cannot say the same of the new generalized interpretation which Einstein, in the framework of his special theory of relativity, gave to the phenomena of gravity, time, and space, so that it might become familiar to a larger public.

By now, the concepts Einstein developed on the structure of space and time should actually have become a greater part of our cultural heritage; the same should be said about some of the consequences his ideas have had on our current views of cosmological development. Future generations will count both the special theory of relativity and the general theory of relativity as among the most important scientific discoveries and ideas of the twentieth century. May this book contribute to the incorporation of Einstein's ideas into our more general culture, and not just into the experts' domain of knowledge. At the opening of the Berlin radio exhibit in 1930, Albert

Einstein began his speech with the following words: "Anyone who mindlessly profits from the wonders of science and technology, but has no more understanding of them than a cow's appreciation of botany and the plants that she devours with great pleasure, should be ashamed of himself."

The first few chapters of this book were conceived during a sabbatical semester I spent at the CERN laboratories in Geneva, Switzerland. I thank my colleagues at CERN's Theory Division for their hospitality. The last chapters were written during a visit to the California Institute of Technology in Pasadena, California. I wish to express my appreciation to my colleagues of its Physics Department for the support I was granted and the stimulating discussions I enjoyed. I also want to thank Mrs. Helen Tuck, who helped me find a number of illustrations. Last but not least, thanks are due to Ulrich Petzold, Hanns Polanetz, Jochen Schürken, and Dr. Klaus Stadler of Piper Editions in Germany, for their useful advice in the completion of this book.

THE CURVATURE OF SPACETIME

Chapter 1

Introduction

For anybody who really penetrates its meaning, it
will be hard not to fall under the spell of this theory.

—Albert Einstein[1]

Near the end of the nineteenth century, there were already signs of
the geopolitical changes that would lead to a redefinition of political
structures across Europe after World War I. At that time, a revolu-
tionary reshaping of the natural sciences also took off. The relatively
conservative German physicist **Max Planck*** developed the first
ideas on **quantum theory**; it is well known that these led to a radical reorganization not only of physics but of the foundations of all
the natural sciences. Shortly after the beginning of the new century,
a young civil servant in the Swiss patent office at Bern, Albert Einstein, gave a new interpretation to the concepts of space and time
that had been cast in stone ever since the Middle Ages, when he
defined the basics of the theory of relativity in 1905. More specifically, it is what we call the "special theory of relativity" that sets new
rules.

In the late nineteenth century, classical physics was seen as the
very model for the natural sciences. It was dominated by Isaac Newton's classical mechanics. The laws of mechanics were interpreted as
unshakable laws of nature; their validity was unquestioned irrespective of whether an investigation concerned the motion of rigid
bodies on Earth or the orbits of planets and stars in the universe.
The cosmos was seen as a gigantic mechanical clock, the motions of
which were determined by classical physics. The bedrock of New-

---

*Throughout this book, bold type indicates the first textual use of a term that can be
found in the glossary.

tonian mechanics was the stability and immutability of space, of the passage of time, and of matter in the universe.

These concepts got a new meaning with Einstein's special theory of relativity; this theory, indeed, had surprising consequences: space and time proved to be phenomena that are dependent on the state of the observer. Also, it implied that there is no universality of mass, a mainstay of Newtonian physics that had to be abandoned; **mass** can, under certain conditions, change into **energy** and vice versa, according to Einstein's equation $E = mc^2$, which in itself is a consequence of the special theory of relativity. This relation says that any amount of matter corresponds to some enormous amount of energy, and that this energy can be calculated by multiplying the mass with the square of the velocity of light, c (c = 300,000 kilometers per second, or 186,000 miles per second). This energy, however, can be set free only in special cases, as in nuclear reactions or stellar explosions in the universe, such as in **supernovae.**

One of the more interesting consequences of the special theory of relativity, which was drafted by Einstein in 1905, is the unification of space and time. It proved impossible to view space and time as separate phenomena, the way Newton had once taught. In Einstein's theory, space and time fuse into one phenomenon which we call spacetime. One consequence of this unity is the dependence of the flow of time on the state of motion: in a rapidly moving system, time flows more slowly than in a system at rest. This is illustrated by the so-called twin paradox: twins, one of whom is at rest on Earth, the other in rapid motion across space—say in a spaceship—will age at different rates.

This is one of several consequences of the theory of relativity that appear to run counter to our experience, to our intuitive notions about the space around us and the seemingly universal flow of time; both of these notions have been developed from birth by all humans. That is why we often speak of a radical subversion of the concepts of space and time by the theory of relativity. But this is not really so; rather, we are dealing mainly with the generalization of these notions that will apply to situations beyond what we experience in daily life. They do apply, however, to processes that involve objects that move at velocities close to the speed of light. (Recall that the speed of light is c = 300,000 km per second, or 186,000 miles per second.)

There is one aspect the special theory of relativity has in common with Newtonian teachings on space and time: according to Newton, space and time are the stage on which dynamical processes of the world are realized. Nothing is able to influence the structure of space and the flow of time. In the special theory of relativity, space and time are replaced by a unified spacetime. But even in the latter context, we are dealing with an ironcast framework which is immutable and cannot be influenced by external constraints.

In the fall of 1907, two years after his first work on the special theory of relativity, Einstein tried to gain a deeper understanding of the phenomenon of **gravitation** on the basis of his new concepts of space and time. He soon realized that he had to expand his theory of relativity so that he could incorporate concepts of gravitation.

Gravitation had been discovered as a universal force of physics in the year 1666 by Isaac Newton, then twenty-three, in his hometown of Woolsthorpe. Legend tells us that an apple falling from a tree in his garden gave Newton the idea that the same force that pulls the apple down to Earth also keeps the Moon moving in its orbit around the Earth, and keeps the Earth in its annual orbits about the Sun. From these ideas issued the Newtonian law of gravity: every massive body exerts an attractive force on every other such body—the more so, the larger the mass of that body.

In 1687, Newton presented his theory of gravitation in his major work *Philosophiae Naturalis Principia Mathematica* (Mathematical principles of natural philosophy), usually called the *Principia*. With this work, Newton laid down the laws of the physical sciences, particularly classical mechanics. In the preface, Newton explains his methods for the description of physical phenomena:

To discover the forces of nature from the phenomena of motions and then to demonstrate the other phenomena from these forces.

The three centuries that have elapsed since the appearance of his *Principia* bear witness to the amazing success of Newton's research method.

Newton deals with gravitation in the third book of the *Principia*, which is titled: "About the World System." It deals mostly with astronomical phenomena. Here, he gives his famous explanation of planetary motion about the Sun on the basis of universal mass attraction, i.e., of gravitation. The mechanics of celestial bodies enabled Newton to define and explain all details of planetary motion. In particular, he was able to deduce why the planets move in elliptical orbits about the Sun, as discovered by **Johannes Kepler** some time earlier. In this way, he impressed his contemporaries in a way no other scientist has been able to do, and only Einstein has recently equaled that feat.

A remarkable success of Newton's theory manifested itself more than one hundred years after his death. It appeared impossible to describe the orbit of the planet Uranus in terms of his theory of gravitation. While measuring the exact orbit of Uranus, small deviations from the theory were detected. Some astronomers and physicists were ready to modify the laws of gravity. But in 1846 the astronomers Urbain Jean Joseph Le Verrier and John Couch Adams independently noticed that the irregularities of the Uranus orbit could be explained in terms of the assumption that there is another planet beyond Uranus, the gravitational effect of which can influence the motion of Uranus. They were even able to determine the position of that new planet with some precision. The same year, the new planet, which was called Neptune, was discovered by the German astronomer Johann Gottfried Galle. Thus, what had been seen as a problem for Newton's theory of gravitation wound up, in the end, its triumphal success.

In 1859, Le Verrier discovered another blemish in Newton's celestial mechanics. This time the planet Mercury, the one that moves on an elliptical orbit closest to the Sun, was noticed not to move in a stationary orbit, as would be expected in the framework of Newton's theory. Rather, its ellipse keeps changing such that its *perihelion*—the point where its orbit is closest to the Sun—slowly revolves around the Sun. Strictly speaking, Mercury's orbit is not a real ellipse; rather, it looks almost like a rosette. Its difference from a stationary ellipse is not very large. Mercury's perihelion changes by only 43 **seconds of arc** per century (to be precise, by 43.11 seconds of arc). Notwithstanding the minuscule size of this effect, the perihelion rotation of Mercury presents a serious problem for Newton's theory of gravitation. An explanation based on the existence of a new planet was eliminated here, since there are only two planets close to the Earth's orbit: Mercury and Venus. Not until the twentieth century was it discovered that the orbit of Venus also shows a very small rotation of its perihelion—as little as 8.4 seconds of arc per century.

Now back to Einstein and the theory of relativity. He needed almost eight years to incorporate gravity into his framework of relativity. His first ideas on gravitation were formulated in 1907, while he was a civil servant at the patent office in Bern, Switzerland. Later, when he was a university professor in Zurich, Prague, and then again in Zurich, he published a series of papers on gravitation. In 1914 he moved to Berlin and became a member of the Prussian Academy of Sciences. The final theory was presented by Einstein to the Berlin Academy on November 25, 1915. It was not just a supplement to his 1905 theory of relativity; rather, this was a fundamental extension, incorporating a novel view of the structure of space and time. It was called, quite rightfully, the "general theory of relativity."

While his old theory, now called the special theory of relativity, was based mainly on the unity of space and time, the basic new idea was to incorporate matter into the unity of space and time. Matter, according to Einstein's hypothesis, cannot be seen without regard to space and time. Rather, it is capable of deforming the structure of space and time. It can induce a warping, or what we will here call a "curvature" of space and time. The result of this deformation is, specifically, the gravitational force. In Einstein's interpretation, it is not a force pure and simple like the electrical attraction between two oppositely charged metal spheres; rather, it is merely a consequence of the geometry of spacetime. An apple will fall, if we follow Einstein, from the tree down to Earth, not because it is attracted by the Earth; rather, the presence of the Earth changes the structure of space and time such that the apple, as it detaches from the tree, has no choice but to follow the given deformation of spacetime. It falls to the Earth without having a choice—just like a train that cannot possibly move outside the predestined path that follows its tracks.

With his idea of the curvature of space and time, or simply of spacetime, Einstein opened new territory for physics in 1915. In all fairness, it should be mentioned that several mathematicians had set the stage for this important step as early as the first half of the nineteenth century.

Einstein's idea meant a further step in a development that had its roots in antiquity. It had been widely conjectured that the surface of the Earth is like a disc, the outer margins of which humankind would do well not to approach. It was not before the late Middle Ages that the circumnavigation of the world then known showed convincingly that the surface is better seen as approaching the surface of a sphere; that it has curvature but not margins. Still, it was assumed that the geometry of our world is determined by laws drawn up by Greek philosophers and mathematicians in antiquity. In the *Elements* by **Euclid** (about 325 B.C.), these laws were systematically formulated, and for more than two thousand years these ideas formed the basis for the teaching of mathematics. Next to the Bible, the *Elements* was the most widely known work in the occidental history of thought.

Euclid's geometry is the geometry of our immediate experience. An area is flat, it has no visible curvature or deformation. That is also true for three-dimensional space: it is flat, it has no inner structure. It is space without further characteristics. A straight line in this unlimited space is just what, naively, we would expect: it runs on to infinity. Two straight lines in parallel will never cross. A circle has a circumference whose length is (2π) times the radius.

In the general theory of relativity, space is not the same as what the laws of Euclidean geometry would make it; rather, it is non-Euclidean: it is deformed. It has an inner structure that is closely tied to the dynamical laws of physics. While **Euclidean space** had been a structureless three-dimensional expanse, it now became a physical medium with its own dynamics that is defined by matter.

When he created his general theory of relativity, Einstein again showed himself to be a master in the critical analysis of new concepts, just as he had done in 1905 for the special theory of relativity. In 1916 he defined his methods of research in a way that reads like an invitation to disobedience:

Concepts that have proven useful when putting order to our observations may easily wind up looking authoritative: we might then forget whence they came, and might accept them as ultimate truths. They become branded as "habits of thought," as a given, an *a priori*, etc. The truth of scientific progress will often be littered by such errors. Therefore, it is not idle gimmickry if we train ourselves in analyzing familiar concepts, looking into the sources and circumstances of their usefulness, their origins in our experience. Such analysis will relativize their authority. They may have to be abandoned if they cannot be fully legitimized. They may have to be corrected if their deduction from observation was on the sloppy side. We may have to replace them by others if there is a new system that we find preferable for whatever reason.[2]

We get another glimpse at Einstein's thinking during the formulation of his theory of gravitation from a note he sent to a colleague at that time:

When I ask myself why it was I who developed the theory of relativity, I found it must be due to one particular fact: a grown-up does not think about problems of space and time. Everything that has to be thought about in this respect, he has thought about when he was a child. I, on the other hand, developed so slowly that I began to wonder about space and time only when I was an adult. It is only natural that I then delved into its problematics much more deeply than a normal child would do![3]

Einstein's theory of gravitation was not only a new interpretation of the mutual attraction of masses, the way Newton had introduced it. It differs from Newton's theory in a number of consequences. One of them is the rotation of the perihelion of planetary orbits. According to Einstein's theory, the space around the Sun is being deformed a bit, by the Sun's gravitation. One result of that space's curvature is a small deviation of the planetary orbits from an exact elliptical form, expressed in terms of a rotation of the perihelion of the orbit. This effect is significant only close to the Sun; and it shows up only in the planets close to it. The planets that are farther removed barely show an observable effect. For the rotation of the orbit of Mercury, Einstein calculated a value that is in good agreement with observation. Today's calculation gives a value of 43.03 seconds of arc and agrees well with Einstein's theory. For the orbit of Venus, the calculation gives 8.6 seconds of arc, and again agrees with the theory.

The orbit of our Earth will also be influenced by the curvature of space around the Sun as Einstein predicted. We get a value of 3,8 seconds of arc per hundred years for the rotation of its perihelion. Observation yields 5.0 seconds of arc; its deviation from Einstein's theory can be explained through the influence on the Earth's orbit by the other planets, especially Mercury and Venus. Einstein's theory makes us expect that not only massive bodies are influenced by gravitation simply because their motion has to follow the curvature of spacetime— even the propagation of light itself will be influenced by the distortion of spacetime. Thus, light rays will be diverted by gravitational **fields**— for instance, by the gravity of the sun. The light of a distant star, when passing close by the solar surface on its way to Earth, will be deflected, according to Einstein's theory, by about 1.7 seconds of arc.

The investigation of the deflection of light by the gravitational effect of the Sun was performed by two English research teams sent by the Royal Greenwich Observatory to two different locations in the tropical zone, where the solar eclipse of May 29, 1919, was expected to have its maximal effects. The results were announced on November 6, 1919, at a special meeting of the Royal Society—the world's oldest scientific society—and by the Royal Astronomical Society in London. Both research teams found a value which, within the unavoidable observational error, were compatible with Einstein's prediction of 1.74 seconds of arc. (It was later noticed that the astronomers made a faulty estimate of the observational error. In fact, the error was of the same size as the predicted effect; in hindsight, therefore, the confirmation of Einstein's theory was not yet compelling. Subsequent measurements of this light deflection, however, established good agreement between experiment and theory.)

Up to November 1919, Einstein had been known only to experts in the field. That changed precipitously when the results of the investigation were published and the deflection of light—as predicted in Einstein's theory—became worldwide news. Overnight, Einstein became the most famous scientist alive. His popularity continues to this day, although only a small group of experts fully understand his research results and their consequences.

The main goal of this book is to acquaint the reader with at least the basic ideas of the general theory of relativity and of its most important consequences, especially their bearing on astrophysics and cosmology. In the present volume, I followed the example of my earlier book about the special theory of relativity, *An Equation That Changed the World*: I have put it in the form of imagined conversations between Isaac Newton, Albert Einstein, and a third person, the fictitious physics professor from the University of Bern, whom I call Adrian Haller. As in the previous book, such dialogues have to be freely invented. Obviously, the parties involved never met, and they could not have met. Newton and Einstein, as they appear in the book, should not be identified too closely with their historical personae. The scenes in the book are located in places that marked important stations in Einstein's life. The actions and utterances that I ascribe to Newton and Einstein are such that they could well choose to make them today, if they were able to comment on what is known about physics at the present time.

I use the form of dialogue (as I did in my book about the special theory of relativity) because it permits a lively confrontation of opinions. The problems conveying the ideas of the general theory of relativity are mainly conceptual. In these conversations, the reader will often identify with Newton; at first, he refuses to accept the ideas of his partners. After detailed discussions, he is quite taken by the new and surprising insights.

Albert Einstein, the way he shows up as Isaac Newton's partner in discussions, is the scientist as he was in 1930, at the age of fifty-one. That was the first summer he spent in his summer house, which he had built only the year before in Caputh, close to Berlin. (The village Caputh lies at the juncture of Lake Templin and Lake Schwielow; both are fed by the Havel River). The drawing up of the general the-

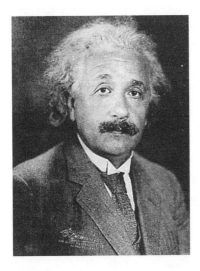

FIGURE 1.1 Albert Einstein in 1930.
(Albert Einstein Archives, Jerusalem)

FIGURE 1.2 Isaac Newton. (Painting by Jean-Leon Huens)

ory of relativity, formulated fourteen years before, had by this time enjoyed its first confirmations by astronomical observations. The discovery of the redshift of distant **galaxies** by **Edwin Hubble** and his collaborators, pointing out the expansion of the universe, had just been made.

Isaac Newton, his discussion partner, is supposed to impersonate the historical Isaac Newton a short time after the publication of his main work, the *Principia*, in 1687. Newton was then forty-five. He appears as a discriminating but conciliatory discussion partner for Einstein and Haller. I have made an effort not to describe the Newton in these scenes too closely in terms of the historical personality of the great physicist. That, had I done so, would have forced me to include some of Newton's rather disagreeable actual characteristics, which would have unnecessarily strained the discussion, without adding to the argument at hand.

The third man in the discussion is, again, the fictitious Swiss physicist Adrian Haller, chosen to represent the ideas of today's physicists and astronomers. I used the Swiss natural scientist, philosopher, and author Albrecht von Haller (1708–1777) as his model. He was called in 1736 to Göttingen, Germany, as a professor of medicine and botany; in 1753 he returned to his hometown of

Bern. He was one of the last universal scholars. In *An Equation That Changed the World*, Adrian Haller "meets" the great English physicist Isaac Newton during a stay in Cambridge—Newton's place of work. Together they travel to Bern, where they meet a thirty-year-old Albert Einstein. Later, the three physicists visit the **CERN** laboratories outside Geneva.

The conversations of the present book, by contrast, start in Caputh, in Albert Einstein's summer lodge, which exists to this day. The protagonists Einstein, Newton, and Haller meet in order to talk about the foundations of the general theory of relativity. They start out with the fundamental concepts of space, time, and matter. These discussions also touch on today's research in elementary particle physics and cosmology. They include present-day attempts to find the reason why most **elementary particles** have mass—which is about to be investigated with a new Geneva accelerator called **LHC** (Large Hadron Collider). Mass, after all, is the basis for all gravitation in the universe, and it therefore defines the large-scale structure of the universe.

After Newton (and hopefully the reader) has been familiarized with the foundation of Einstein's theory of gravitation by Haller and Einstein, and after he has changed into a convinced adept of Einstein's ideas, the scene of action moves to another place. In the second part of the book, the threesome of, by now, inseparable physicists winds up in Pasadena, outside Los Angeles. This is where Edwin Hubble discovered the expansion of the universe in the late 1920s. Both Haller and Einstein are well acquainted with Caltech (California Institute of Technology) and with the observatories it operates. Einstein had been a frequent guest there in the second half of the 1920s. Here the discussion touches on the knowledge of today's astrophysics and cosmology: **black holes**, gravitational waves, and whatever happened right after the Big Bang, some twelve billion years ago.

Different aspects of the special theory of relativity are raised in this book. I have tried to cover these here so that the reader need not have read my earlier volume on the subject and, if unfamiliar with its main ideas, will not be lost. (I briefly sketch the ideas of the special theory of relativity in an appendix to this book.) *The Curvature of Spacetime* should not be seen as simply an extension of *An Equation That Changed the World*, although the same people are protagonists in this book, and although the general theory of relativity is, historically, a further development of the special theory of relativity. Still, the general theory of relativity has only tenuous connections with the special theory of relativity and is, in fact, an independent theory of gravitational phenomena. On the contrary, the special theory of relativity may be seen as "nothing but" a new framework for a modern description of well-established physical phenomena and forces.

CHAPTER 2

Meeting Einstein and Newton in Caputh

Behold the stars, behold how they teach
how far the master's prowess will reach.
Newton, by rights, can claim paternity
for telling the orbits they run in eternity.
—Albert Einstein[1]

At 7:00 P.M. sharp, Crossair flight 1215 started from Geneva's airport; one and a half hours later, it landed at Tempelhof Airport in Berlin. Adrian Haller looked pensively around the airfield which, although fairly small by today's standards, had played an eventful role in the history of the German capital city from June 1948 until May 1949. It was then the only link to the western part of Germany, after Stalin had ordered the blockade of West Berlin. The monument to the airlift (erected later in front of the terminal building) reminds us of that oppressive time; after all, it marked one of the climaxes of the Cold War.

Haller was following an invitation Humboldt University and Potsdam Institute had extended to him. He took the subway from Tempelhof Airport to its station by the Zoo, in the western part of town. From here, he was able to take the light rail to Caputh, a small town outside Potsdam, where Lake Templin and Lake Schwielow meet. Close by, there was the lodge which was to host him, courtesy of the institute.

The cab driver had a hard time finding the address. He would have had a much easier time in the early 1930s, Haller thought to himself. At that time, after all, the lodge was one of the best-known addresses in the Berlin area, to scientists. Albert Einstein and his second wife, Elsa, had moved into the lodge that he had built for himself. Made of nothing but wood, and in a design that was modern for its time, it attracted a steady stream of illustrious visitors, that came flocking to Einstein's abode, to the astonishment of the villagers. Among the visitors, there was the Indian poet and philosopher Rabindranath Tagore, with flowing white robes, surrounded by followers who came for philosophical discussions with Einstein. Besides numerous physics colleagues, artists like the painter and graphic designer Käthe Kollwitz, the painter Max Liebermann, and the novelists Gerhart Hauptmann and Heinrich Mann, they all paid their calls on Einstein.

Einstein, an enthusiastic sailor, kept a dinghy close to the house. He developed a fondness for country life in this location; it was, in fact, hard to locate him in his city apartment in the center of Berlin. Unfortunately, he was not to enjoy the amenities of his new country home for long. Anti-Semitism was on the rise in Germany; and as it became clear that the National Socialist Party was to assume power, he decided to leave Berlin and his country of birth, Germany. On December 10, 1932, he and his wife left for Belgium; the following year, they traveled to New York on board the American steamer *Westernland*. He was not ever to see his beloved country lodge again. It was confiscated by the National Socialists in 1934 to house officers of the Luftwaffe during World War II. After the war, it became a refugee shelter. In 1979 it was transformed into a meeting place for scientists. After German reunification, it passed into the hands of the Hebrew University in Jerusalem.

FIGURE 2.1 Einstein at the window of his summer house in Caputh near Potsdam around 1930. (Albert Einstein Archives, Jerusalem)

After numerous telephone conversations, Haller arranged with his colleagues in Potsdam to be lodged in Einstein's summer house during his visit. It had not been damaged in the war; during the decades of the Communist East German state, it served as a memorial to Einstein and was administered by the Academy of Sciences. A neighbor who tended the house showed Haller around. After he chose a bedroom that had been made up on the first floor, he proceeded to take a look around.

Built basically as a prefabricated structure after a design of the architect Konrad Wachsmann, and erected in 1929, it is a low-strung building with a large living room, a kitchen, and a big terrace, joined to a two-story structure with five rooms and a single bathroom. The larger of the rooms on the ground floor had served as Einstein's office and bedroom. The French-style, full-length windows had been a special whim of the original owner, just as the fireplace built out of white marble in the living room.

Haller retired to his bedroom soon after ten. He was going to be fetched at nine the next morning. He set his alarm clock for eight and went right to sleep but was abruptly awakened by voices familiar to him. Could it be true? He jumped out of bed, got dressed, and ran downstairs to the living room from where the voices had reached him. Slowly he opened the door and took a peek inside.

"Well there you are, Professor Haller, welcome to Caputh!" was the greeting he heard. There, in fact, was Albert Einstein, sitting in

FIGURE 2.2 Exterior view of the south side of Einstein's summer house in 1930. (*Photo:* Konrad Wachsmann)

FIGURE 2.3 Einstein's summer house in 1995.

front of his fireplace, alive, pipe in hand. He got up to welcome Haller.

"It has been a few months since we last met in Geneva.* At the time when we bid our farewells then, I did not want to tell you; but I knew that we would see each other again soon. You'll not be astonished that our good friend Sir Isaac is again with us. He got here yesterday morning, and we spent the day enjoying a little sailboat outing on Lake Templin."

And indeed, the second person had gotten up and extended his hand in greeting. Haller was barely surprised to find himself opposite Isaac Newton, at about forty-five years of age and looking much the same as when he had said good-bye to him at Geneva's airport several months ago. Einstein, in comparison, had aged some. His longish gray hair framed his wide-eyed face. He seemed to interpret Haller's attentive glance.

"Newton and you, Haller, look the same as when I left you. The time machine appears to have been busy only on me. True enough,

---

*See Harald Fritzsch, *An Equation That Changed the World: Newton, Einstein, and the Theory of Gravity* (1994).

I am fifty-one now, a good twenty years older than when last we met. But that again has its merits because I have been able to get a lot done. My theory of gravitation, the general theory of relativity, has developed from a cheeky lass to an attractive young lady that looks like she might be going places. In fact, my old theory of relativity that we chewed over at length when last we met is now coming into its own. But why don't you first have a seat, my dear colleague? I can imagine it will be a long night for you. If I see things correctly, Newton will poke holes into our stomachs with his curious questions about the fate of our understanding of gravitation.

Haller had already given up all hope of getting back to bed. He dropped into an easy chair by the fireplace. Newton brought in a piping-hot pot of fresh coffee from the kitchen and sat down next to him. He looked at Haller and started out:

"Again, be welcomed to our old Olympic Academy, now complete, my dear Haller. I suggest we continue in the same manner as before. Einstein already gave you the hint. During our discussions a while ago, we kept gravity pretty much out of the picture."

EINSTEIN: And a good thing it was. After all, at age thirty-one I had no idea, or maybe just the smallest inkling, about how to crack this nut.

NEWTON: Easy, dear Einstein, I am not so sure that you have actually cracked the nut in the meantime. As a matter of fact, I'm not sure there is a nut to crack in the first place. I suggest, for starters, that I say a few general words concerning the problem at hand.

EINSTEIN: Shoot, Sir Isaac. I can't wait to find out. During my time in Zurich, when I read your *Principia*, I had the idea that you racked your brain about gravity much more than you're willing to admit.

NEWTON: Allow me first to make a few general remarks on space and time. We know that space and time exist independently of matter in the universe. They are, so to speak, the receptacle in which matter is embedded. Space has three dimensions, time has one. That means that we can always define a position in space with three numbers, the three coordinates; while one number is enough to fix time uniquely.

EINSTEIN: What you just said is valid for your mechanics as well as for my theory of relativity. But I would like to point out right now that we will soon have problems with your comment that space and time exist independently of matter. That is the point where my general theory of relativity deviates from what you said.

NEWTON: You seriously mean to say that matter could influence how space is structured and how time flows? . . .

He stopped suddenly, murmured an excuse, and left the room.

EINSTEIN (*chewing on his pipe*): I was sure Newton would take my bait.

HALLER: You mean Newton secretly played with the idea that matter could influence space and time, which is the case in your general theory of relativity?

EINSTEIN: Why don't you read his *Principia* attentively? I mean, read between its lines, too? I am pretty certain he had at least an inkling in that direction, but not much more. Not bad, considering that it was three hundred years ago.

NEWTON (*returning*): Excuse my brief absence; I needed a few minutes to put my thoughts in order. Let's go on. Space and time are homogeneous; that means, there are no points in space that stand out, no moments in time that do likewise. This implies that space is infinite and that the flow of time in our cosmos has always been as we see it now, and will remain so forever. Any coordinate system that we use to describe space can be moved around arbitrarily. All such systems are equally valid. Space and time have an entirely democratic structure—if you allow me to use this political term.

HALLER: You can spin this yarn further. A coordinate system in space can be rotated any which way. The directions of the individual coordinates are in no way fixed. No direction stands out in space, they are equivalent. Space is isotropic.

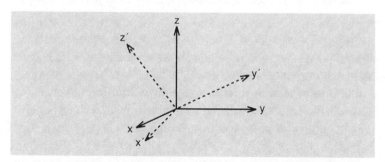

FIGURE 2.4 Three-dimensional space is spanned by three coordinate axes that are at right angles to each other. They can be arbitrarily rotated.

NEWTON: I'll grant that. Space is homogeneous and isotropic, and time runs on homogeneously. But now to matter as it moves. A material object or, for simplicity, a massive point—less abstractly, a small iron ball—will move through the universe in a straight line unless it is subject to some force, whichever force that may be. We can always find a coordinate system that moves at the same speed as our sphere itself. In that system, the ball is at rest. This means that it doesn't matter if I observe the motion of a body in space from a moving coordinate system or one at rest, as long as the moving system has the same velocity across space. All these systems are equivalent from the physical point of view.

EINSTEIN: That also would mean that there is no absolute motion in the universe. All motion is relative; it is dependent on the reference system at hand. Things do not move absolutely, they move with reference to other objects or locations.

HALLER: We did discuss this before. Let me remind you that, in this manner, we introduced **inertial systems**. We defined these as systems of reference in which a massive point can move freely in a straight line and with constant velocity. That is, of course, an idealized limiting case. A massive object—say, a spaceship—will not, in general, move freely across the universe. It is subject to the mutual attraction of masses, and thereby to the gravitational force of celestial bodies.

NEWTON: Basically, any inertial system is a figment of the imagination of physicists: we can live with it; above all, we can use it for calculations. But it does not exist in the real world. I always thought it peculiar that the **acceleration** of a body in an inertial system has a meaning that does not depend on the reference system; conversely, its velocity, which can be defined only relative to a reference system, is not absolute. *To repeat: velocity is relative, acceleration is absolute.*

HALLER: Precisely. This means that we see some pretty peculiar phenomena inside an accelerated reference system, say, inside a car that has a rapidly increasing velocity. The driver is being pushed back into his seat, objects that are not fastened down will fly back inside the car, etc. In a word, forces that do not exist in an inertial system show up in an accelerated reference system. Engineers and physicists call these forces inertial forces; they are a consequence of the inertia of material bodies. They will not participate willingly in the accelerated motion the system is forcing on them. Rather, they want to keep moving as before. They resist acceleration. And the result is a force that the reference system (in our example, the car) applies to our material probe so that it will participate in the acceleration. This force increases as the mass of a body increases. And I think we'll have to pay closer attention to what this force does. Am I right, Professor Einstein?

EINSTEIN (*looking at his watch*): Yes, you can say that again. For years I pondered over this question until I found "the real McCoy"—but we won't be able to cover all of that today.

HALLER: I second that. For my part, I just finished my trip here from Switzerland, and I don't think I feel up to a useful discussion that can clear up this problem. I will retire, and I suggest that tomorrow morning we pick up where we are leaving it now.

EINSTEIN: Good idea. Sleep well in my summer house. I find it wonderfully peaceful, just as it was in the 1930s. Gentlemen, until tomorrow morning.

With that the trio dispersed. Einstein and Newton did as Haller, and retired to their respective rooms.

CHAPTER 3

"Subtle is the Lord"

The Lord God is subtle, but malicious he is not. . . .
Nature by her very grandeur hides secrets, but not
by ruse.

— Albert Einstein[1]

Haller got up shortly after seven. The clatter of dishes coming from the kitchen had awakened him. Breakfast was being prepared for the guests. Half an hour later, Einstein and Haller sat at the table in the living room. Einstein reflected on the many pleasant days he had spent here long ago.

"The best thing was sailing, especially when everybody was trying to get hold of me. The most enjoyable moments for me were when the wind died down so that my boat drifted aimlessly, and I could not be reached. Then I had all the time in the world to think."

"So why don't we have one of our discussions while sailing?" said Newton, appearing in the doorway.

"Let's make a note of it. But why don't you sit down right now? We can get an early start while having breakfast. Shoot, Sir Isaac— today, we're talking about space, time, and gravity."

NEWTON: I recently read that you claim it is possible to understand gravitation with the aid of your theory of relativity, Einstein. One should probably say, "understand spacetime and gravitation jointly"; after all, space and time are joined together as one in the framework of your theory of relativity. I assume you don't mean to rescind that.

EINSTEIN: As you wish, we can call it spacetime, or even timespace, but that is not of fundamental importance. What does matter is that space, time, and gravitation are closely linked together so that we can view these three aspects of the universe as different properties of one and the same phenomenon.

FIGURE 3.1 Einstein at the helm of his boat, *Tümmler*. (*Photo:* Hermann Landshoff, 1930)

Newton got up, shaking his head. He gave Einstein a disapproving look: "Not so fast, dear colleague. In your theory of relativity—more precisely, as we say today, your special theory of relativity—you said that an absolute space and an absolute time do not exist in our universe; in fact, that they cannot exist. Rather, the flow of time and the characteristics of space depend on the reference system; they depend on the observer. Two different observers who are in a state of uniform motion with respect to each other, observe two **events** in space in different ways. For instance, they discover that the difference in time between two events is not the same for both of them. I have accepted all that even though, as you know, it cost me great difficulty and a conscious effort to throw my ideas of absolute space and absolute time overboard. I also know that the theory of relativity casts a new light on the forces of nature. For instance, we cannot look at the electrical and magnetic forces as separate phenomena, as was done in the early days of the teachings on electricity. Relativistically speaking, there is only a single force, which we call electromagnetism."

HALLER: Right you are. An electric field of force—say, the field created by a metal sphere that is electrically charged—is no longer electric only in character as soon as it is being observed from a moving reference system. It is now accompanied by a magnetic field.

NEWTON: Yes, I have got to that point myself. An electric current in a wire will create a magnetic field of force around the wire; it is

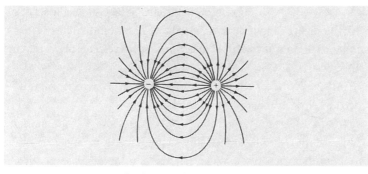

FIGURE 3.2 Two oppositely charged objects attract each other. This is due to the electrical fields that surround the charged objects. The figure shows the shape of the field lines. The two objects also attract each other due to gravity. This attraction is weaker by many orders of magnitude. Both forces have one thing in common: their strength decreases with the square of the distance.

due to the fact that the electrical current consists of many **electrons** that move in the wire. This motion produces a magnetic field. And this is ultimately a consequence of the theory of relativity. What I mean is this: electric and magnetic fields are fields of force, given by Nature. For starters, they have nothing to do with the theory of relativity. Physicists in the last century worked well with electric and magnetic phenomena even though they had no inkling of the theory of relativity. Let's look at two electrically charged spheres, one with a positive charge, the other with a negative one. Opposite charges attract each other; that means there is an attractive force acting between the two spheres. Its strength depends on the distance between the two spheres. The larger the distance, the weaker the force—it decreases with the square of the distance. If I double the distance, the force decreases by a factor of four. This simple law of the strength of the field is, of course, not a direct consequence of the theory of relativity. Still, our theory does cast a new and interesting light on it. The same thing might be true for the gravitational force.

But let's stick with our two spheres: if we remove the charges from both, there will no longer be any electrical attraction between them. But gravity remains—that means the masses of both spheres will still attract each other. This gravitational force increases with the mass of the metal sphere. Also, the mutual attraction of masses decreases with the distance between them. To be precise, it decreases with the square of that distance. My law of gravitational force is very similar to the law of electrical force: just replace electrical charges by masses . . .

HALLER: However, the gravitational constant G, which you introduced in physics, must play its part here.

NEWTON: Certainly, the knowledge of mass alone is not enough. My gravitational constant simply expresses the strength of mass attraction between two bodies that are at a meter's distance from each other, and that have a mass of one kilogram each. Of course, in my time, this constant was not very precisely known. Tell me, what is its numerical value today?

HALLER: In comparison with other constants of nature, Newton's gravitational constant is still not all that precisely known. Today's value is 6.67259 times $10^{-11} m^3 kg^{-1} s^{-2}$. That means, the mass attraction between the two bodies we talked about causes a mutual acceleration in the amount of 6.67259 times $10^{-11} ms^{-2}$. In one second, the relative velocity of these bodies increases by an amount of no more than a 60 trillionth of one meter per second. In comparison, the increase in velocity of the same bodies in the Earth's gravitational field is almost 10 meters per second.

NEWTON: I would like to remind you that the electrical force is a consequence of the existence of an electrical field of force, which surrounds every body that is electrically charged; this body, however, has its own space structure: it is a property of space that surrounds the charges. It is, if you wish, a property of the space. Any given point in space is, therefore, endowed not only with its coordinates in space and time but also with the strength of the electrical field in its location. A volume that is filled with electrical fields is richer in structure—we might say, it is less empty—than a volume in the absence of a field.

Things might be very similar when we talk about gravity. A body is surrounded by its gravitational field. I read in some book that, in the last century, gravity and electricity were used to speculate about the nature of electrical force. That means people speculated about a similar behavior of electrical and gravitational forces before there was experimental evidence that both decrease with the square of distances. To me, that means: electrical and gravitational fields show an analogous behavior.

HALLER: Not quite—don't forget the fact that the attraction between masses in general is much weaker than the electrical forces; in fact, it is tiny by comparison. It depends, of course, on the charges. If we compare the electrical attraction between a **proton** and an electron in a hydrogen **atom** with the gravitational attraction between the two of them, we see that mass attraction is weaker than electrical attraction by a factor of 10^{38}. This is a huge number; to put it in words, it's one hundred times a billion times a billion times a billion, and again times a billion. In other words, you can forget about gravitation when you compare it with an electrical force.

Macroscopically, this is obviously not the case. The point is that a macroscopic body—say, the cup I am holding in my hand—is electrically neutral as seen from the outside. Even an electrically charged macroscopic object has an external charge that is minuscule when compared to the many charges it holds on the atomic level inside its volume. If this were not so, the electrical forces would be enormously strong. Take a nucleus of uranium; it holds 92 protons, each of which has a positive electrical charge. They repel each other so strongly, as equal charges do, that they almost cancel the strong **nuclear force** that binds them together. That is why a minor collision of a particle with this nucleus can cause it to split up. This is the very process that is exploited for energy production in a nuclear reactor.

The fact that the electrical forces are so much stronger than gravity might be seen as a hint that gravitation is even qualitatively a phenomenon quite different from electricity. This very idea is realized in Einstein's theory of gravitation, as we are about to find out.

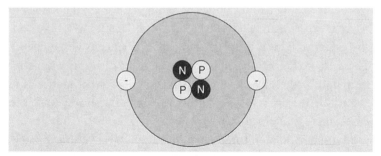

FIGURE 3.3 Schematic sketch of a helium atom. Its nucleus consists of two protons and two neutrons, the shell of two electrons. The atom as a whole is electrically neutral.

EINSTEIN: One more thing, Sir Isaac. There is an important difference between the electric and gravitational forces: electrical forces can be either attractive or repulsive. This depends on the charges in each case, whether they are equal or opposite. This is a difference that does not exist in gravitation. Masses *always* attract each other. There is no such thing as gravitational repulsion or, if you wish, antigravitation, no matter what some science fiction books might cook up.

NEWTON: Since you bring up the strange concept of antigravitation—I'm not all that sure that it does not exist after all. When we met at CERN a while ago, we talked about **antiparticles**. Every particle in nature has an antiparticle; for the proton, there is the **antiproton**. What about gravitation among antiparticles? It could be that a proton and an antiproton do not attract each other gravitationally, but that they repel each other. In my theory of gravity, antiparticles might appear with negative mass, the analogue of negative electrical charges. That would give us antigravity in addition to the gravity we are accustomed to. Since normal matter consists of particles and not of antiparticles, there would be no changes in the universe for normal gravitational phenomena. But things might change when **antimatter** enters the picture. Two stars, one made up of matter, the other of antimatter, would not attract each other: they would repel each other. That would be some impressive spectacle in the universe.

EINSTEIN: That would certainly surprise me. Anyway, this is impossible in the framework of gravitation the way I go about it.

NEWTON: Your theory, fine as it is, might—if you forgive me—be either wrong, or it might apply only to normal matter. After all, we're not talking about politics or about art; we are talking about physics: experiment will have the last word. That is always the decisive evidence. This is the good thing about science as we practice it. There are no half truths; mushy or misleading concepts will necessarily fall by the wayside. So, Haller, what does experiment tell us about this? Antiparticles will either rise or fall with respect to our Earth, there is no third way.

HALLER: You're hitting on a touchy subject. Unfortunately, physics does not have as clear an answer as you might wish. At the CERN laboratories, which I just visited, there are large amounts of antiprotons with which our colleagues perform experiments. But as

far as gravitation is concerned, we have learned very little; to be hon-
est, we have not learned a thing. There are plans to do some free-fall
experiments with antiprotons, but I don't know if they will be realized in the near future. There are many difficulties: antiprotons do
not survive long when they are in danger of reacting with normal
matter around them. That will lead to their annihilation.

There are indirect theoretical arguments that mitigate against
antigravitation being effective when we deal with antiparticles. We
know that the mass of the proton has several components. One of
them is due to antiparticles—or, more precisely, to antiquarks.
These antiquarks are the antiparticles of the smallest particles—
quarks, introduced in the early 1960s by **Murray Gell-Mann** and
George Zweig at Caltech, as the basic building blocks of all nuclear
particles. The proton is made up mainly of quarks, the antiproton
mainly of antiquarks. But as early as the 1970s, we found that inside
the proton there are also antiquarks. This is a consequence of the
very strong forces between the quarks, and of Einstein's special theory of relativity. For this reason, a small part of the proton's mass—
say, about 10 percent—is due to antiquarks. That means this is mass
due to antimatter. Now we do know, as far as gravitational attraction
is concerned, that we'll have to deal with the entire mass of the proton that enters into your equation. This includes the mass that is due
to the antimatter inside the proton. Therefore, it is basically impossible that antimatter has different gravitational behavior when compared with normal matter.

Still, this is an indirect argument. I would feel better if we could
directly observe that antimatter drops toward the Earth in its gravitational field, just as normal matter does; that it is not being repelled
instead. Nevertheless, we can assume with some certainty that there
is no such thing as a repulsive gravitational force. That's a good thing
for you, Einstein. Gravitation among antiparticles would put an end
to your theory. I would not know how to incorporate this into your
concept; I believe this will become clear pretty soon.

EINSTEIN: In fact, I'm quite sure that you, Newton, will soon find
antigravity as unedifying as I do. **Lichtenberg**, the great experimental physicist at Göttingen University, liked to stress in his lectures
that the laws of electrical forces are quite general; that they can actually be adopted for human behavior. Persons with different
"charges," who usually attract each other, do occasionally repel each

other as soon as their "charges" have become equal. Exchange the word *charges* for *names*, then the marriage of two people gives them equal charges. Thank God, this seems not to be the case for gravitation. To marry gravitationally is therefore more commendable, and more promising, than to marry electrically.

NEWTON: That is a special feature of my law of gravity: there is only attraction if there are positive masses. If there is, indeed, no antigravity—and let's assume that this is so, because Haller's argument about the antiquarks inside the proton sounds plausible if not compelling—then the reason must be that there is only one kind of mass. Mass is always positive, even the mass of antiparticles. But apart from the fact that there is an obviously big difference of strength between gravitation and electricity, the gravitational and electrical forces have strong similarities. Therefore, I believe that the theory of relativity can uncover new, interesting aspects of gravitational phenomena, as it has done for electrical and magnetic phenomena. But there is nothing that is fundamentally novel: electrical charges, as well as masses, exist even in the absence of the theory of relativity.

Therefore, I cannot agree with you, my dear Einstein, that some kind of unity of space, time, and gravitation is important, as you indicated before. I don't see that gravitation is a consequence of the structure of spacetime. Let's face it, electrical phenomena have nothing whatever to do with the structure of spacetime either.

Einstein got up, walked to the window, and pensively looked out toward the lake. After a short while, he turned around, walked toward Newton, looking at him, his large eyes attentive and good-natured.

EINSTEIN: My dear friend, let's try to get this straight. True enough, gravitation is exactly what you just said, a manifestation of the structure of space and time. The gravitational force between massive bodies is, in contrast to the electrical force, not an independent phenomenon of nature. Rather, it is, if you wish, a consequence of the geometry of space and time. This and nothing else is the quintessence of my general theory of relativity, which I completed here in Berlin in 1915.

As Einstein was still holding forth, Newton jumped up and stared at him while pacing across the room. He was clearly excited.

NEWTON: But that sounds absurd. How can you say that a physical force which we sense everywhere, which keeps us on the ground, has something to do with the geometry of spacetime? Space is space and force is force, Einstein. Both are as different as fire and water.

Newton took an apple from a bowl on the table and let it fall down on the wooden floor.

NEWTON: Here, look at it, the apple falls down because the Earth attracts it. It is just a consequence of the law of gravity. You will not tell me, my dear colleague, that this force is not for real; that it is a figment of our imagination. Or worse, a consequence of the structure of space and time. How absurd! You are not seriously claiming that the apple falls down because of some complicated structure of space and time that coerces it to do so? As we know, your theory of relativity is only relevant when the velocity of the object is close to that of the speed of light, c. This is not the case here—the apple falls straight down with a velocity that is ridiculously small in comparison with the speed of light, c. The apple falls at a rate of just a few feet per second, and we, the observers, are fully at rest. In such circumstances, my laws that I wrote down in the *Principia* will be valid. So why all this babbling about the theory of relativity, be it the special or the general version of that theory?

EINSTEIN: My colleagues in Berlin reacted the same way as you are doing right now, Newton, when I first presented my general theory of relativity—that is, my theory of gravitation—to the Berlin Academy on November 15, 1915. Most likely, a few of them thought me raving mad; most were sure I was on the wrong track, and would soon realize my mistake.

NEWTON: All right, let's assume, in your favor, that this is not so. Rather, I believe there is something to your theory, and I suggest we go about it in a systematic fashion. Back to electricity: as we have seen, mass and electrical charges play an analogous role in our scientific game. Nonetheless, this analogy appears to me a bit facile. The way I see it, the mass of a body and its electric charge are two entirely different things. Isn't it true that only certain charged particles carry electricity, while all particles—or almost all of them— carry mass, irrespective of their charges?

HALLER: You are perfectly correct. Every body consists of atoms; the atoms, in turn, are made up of a nucleus and a shell of electrons. The electrons are electrically charged. Every electron carries the

same amount of charge, which we define to be negative. This definition is somewhat arbitrary; it was introduced in the eighteenth century by the American researcher **Benjamin Franklin**. He could just
as well have defined it as positive. In fact, that would make more
sense—but hindsight doesn't help. An important feature is: *the
amount of charge an electron carries is a constant of Nature*; its negative is simply called the elementary electric charge. All electrically
charged objects in nature possess a charge that is an integer multiple
of this elementary charge.

NEWTON: Hmm, I guess the electric charge of the nucleus must
be positive; what then is the connection with the electrons and their
negative charges?

HALLER: Actually, none. Atomic nuclei are made up of protons and
neutrons. The protons are positively charged, the neutrons have no
charge; and that is why we call them neutral. The entire atom is neutral to the outside. Electrically, the positive charge of the nucleus is
balanced by the negative charge of the electrons in the outer shell. For
example, the simplest atom, the hydrogen atom, consists of just one
proton as the nucleus, and only one electron in the shell. The charge
of a proton is the same as the electron charge. There is just one difference: the proton charge is positive, the electron charge is negative.

NEWTON: Now wait a minute. You claim that the nuclear
charge—in the case of hydrogen, the proton charge—has nothing to
do with the charge of the electron?

EINSTEIN: It doesn't. After all, protons and electrons are completely different particles.

NEWTON: Isn't it strange then that their charges are numerically
equal? Suppose their charges are actually just a bit different—say, by
1 percent or so—that would mean the atom would have a net charge,
if only a small one. How precisely do we actually know that the
charges of both particles are equal except for the sign?

HALLER: I have to disappoint you, Sir Isaac. Although you are
quite correct that, in principle, the electric charges could differ a bit
between proton and electron, this would have catastrophic consequences: the atoms that have a net charge, albeit only a small one,
would repel each other. As a consequence, large accumulations of
matter, such as we see in stars and planets, could not even form. If,
in a split second, the proton charge were a little smaller than the
electron charge, the consequences would be, as I said, catastrophic.

All bodies would explode. Our Earth would turn into a huge ball of matter that expands radially in all directions.

NEWTON: Of course, I would have thought of that myself. It is then clear: the stability of matter in our macroscopic world makes it indispensable that proton and electron charges are precisely equal, except for the sign in front.

EINSTEIN: That's odd, and it cannot possibly be a coincidence. Haller, you work also with elementary particles? Couldn't you tell me how a proton knows that its electric charge has to be precisely the same as that of an electron, except for the sign in front? Both particles appear to have little connection to start with, given that the proton actually consists of several quarks, quite in contrast to the electron. They differ from each other, it appears to me, just as apples differ from blueberries.

HALLER: Okay, since you insist on this particular point: protons and electrons have nothing to do with each other for starters. It is therefore legitimate to ask why their electric charges are exactly the same in quantity. This strange property has its own name, which we call the *universality of charges*. More specifically, and maybe in a somewhat stilted fashion, we call it the quantization of the electric charges.

EINSTEIN: To give it a name is one thing; to understand it, another. So let's have it, Haller. Do you know why the charges are numerically equal, or don't you?

HALLER: Since you ask me directly, we don't really know that to this day. It remains a puzzle. Most of the particle researchers assume that protons and electrons are ultimately related, notwithstanding their thoroughly different appearance. The universality of charges is seen as the outer signal for their relation; you might call it the ring that shows them to be married.

EINSTEIN (*turning to Newton*): You see, physicists are still working on the basics. To this day, they don't really know why the hydrogen atom has no net electric charge. Scandalous, isn't it?

HALLER: I'm really sorry I cannot come up with a satisfactory answer. But I can assure you that this is a particularly hard nut that was given us by the Creator to crack.

EINSTEIN: Never mind, Haller. There is something pleasant about not knowing everything. Some time or other we'll find out the true reason. Recall: "Subtle is the Lord," but without malice. It is just too bad that Newton and I cannot take part in the game. But let's leave that where it is, for now. Thank goodness that the problem of the charge is not really important for our understanding of gravitation, at the level we are discussing it.

NEWTON: The electric charge, if any, of an object is where its electric force originates. Today, we know that the source is in the charges of the electrons and protons that make up the object at hand. Similarly, the gravitational force of an object originates in its very mass; for the electron, its mass is at the basis of its gravitational force.

HALLER: We could describe the mass of a particle as the gravitational charge of that particle.

NEWTON: Exactly, there is just one difference: in the case of gravity there is no such basically incomprehensible, or ununderstood, universality of this charge that we just mentioned. True, the electron has a given mass . . .

EINSTEIN: If you express the mass of the electron in energy units of, say, a million **electron volts**, or MeV, following my law on the equality of mass and energy, then the electron mass makes up 0.511 MeV.

NEWTON: The proton, in turn, is exactly 1,836 times heavier than the electron; it has a mass of 938 MeV. That means the gravitational field of the proton exerts a force that is 1,836 times stronger than that of the electron. It is out of the question to speak of any kind of universality of gravitational forces in the way we can do that for electrical forces. The force exerted on a particle depends on its mass; but, since different particles have different masses, the individual amounts of force acting on them are also quite different. What is the origin of this strange phenomenon? More basically: what is the reason for different massive bodies to attract each other?

Einstein had gotten up during Newton's words and had walked over to a bookcase. He took out a volume and opened it.

EINSTEIN: My esteemed colleague, this is just the question I put to myself in 1906; and I did so walking, I should say, in your footsteps. Permit me to quote from your *Principia*. Toward the end of your book, you wrote:

So far I have explained the phenomena of celestial bodies and of the oceans by means of the gravitational force; but not yet have I discussed its origin. Clearly, this force is caused by some phenomenon that penetrates to the very center of the Sun and of the planets without losing any of its effectiveness. It does not act in proportion to the surfaces of those particles it penetrates (as mechanical causes do), but rather in proportion to the amount of mass; and it acts in all directions equally, out to immense distances. Its action decreases with the square of the distance. Gravity acting on the Sun is composed of the action on each of its constituent particles; and again, at increasing distances from the sun, it decreases exactly as the square of the distance from the masses on which it acts. . . . I have not yet been able to get to the point where I could derive from these phenomena the actual origin of the properties of gravity; and I do not make hypotheses.

So you do write that you do not want to make hypotheses; but in this particular respect, my dear Newton, I do not trust you. I am sure that, at the time, you tried out plenty of hypotheses.

NEWTON: It doesn't have to be a secret anymore: true enough, I did hypothesize. But the results were never really satisfactory. If I understood you correctly, your general theory of relativity is not far from guesswork either. The problem I was unable to solve was the fact that the origin of gravitation—if we can use that word factually—is ubiquitous; it penetrates the interior of the Sun, of the planets without trouble. Now that I have learned something about electrical phenomena, this sounds even more peculiar. As you know, electrical energy can be shielded, but gravity penetrates everything, like space itself.

EINSTEIN: I do hope my theory is more than a hypothesis. After all, everybody calls it the theory of gravitation nowadays.

NEWTON: That remains to be seen. As long as I don't know what the basic idea of your theory is, I cannot agree with you.

HALLER: Don't be so skeptical, Newton; in a few days, I'm sure, you'll speak quite differently about Einstein's theory. By the way, you just said something very interesting: you said that the origin of gravity is all-pervasive, just like space itself. Suppose we are able to make the structure of space, or maybe the structure of space and time, responsible for gravitational phenomena; in that case it would be easy to understand why there is no shield against gravity, just as there is no shield against space and time. And that is exactly what Einstein's theory is all about, as we'll see soon enough.

But let's get back to the analogy between mass and charge, which we discussed before. There is a further important difference between these two concepts—one that we have not touched on. In all physical processes, the electrical charge remains unchanged—it is strictly conserved. This is not the case with mass. In many physical processes, particularly processes of nuclear particle physics, there is no conservation for the total mass of the particles or nuclei involved.

NEWTON: You're right—that is indeed an important difference. When we join a proton and a neutron and make an atomic nucleus out of them—to be precise, the nucleus of what we call heavy hydrogen—then, the mass of the newly formed nucleus is a tiny bit smaller than the sum of the masses of the proton and neutron.

HALLER: This is illustrated even more impressively when matter and antimatter annihilate. When we collide an electron and its antiparticle, the **positron**, the two particles vanish in a flash of two **photons**. The original particles in this process had a total mass of about one MeV; after the radiative annihilation, all that is left are the two photons; and those photons do not have a rest mass at all.

NEWTON: Now we are getting into pretty odd territory. An electron-positron system, shortly before annihilation, has a mass of about one MeV. That means it generates a gravitational field all around—no matter that this field is so weak that we can barely detect it. But now the two of them collide, and they radiate off two photons. In the process, all mass is gone. There are just these two photons that vanish with the velocity of light. Doesn't that say that the gravitational field is also gone?

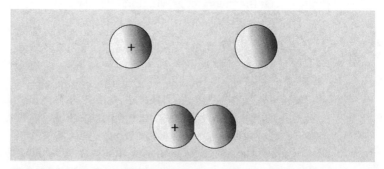

FIGURE 3.4 The strong nuclear force binds a proton and a neutron together to form a *deuteron*, the nucleus of the heavy hydrogen atom. The mass of the bound nucleus is a little less than the sum of the proton and neutron masses.

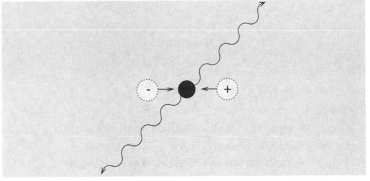

FIGURE 3.5 Annihilation of an electron and a positron into two photons.

EINSTEIN: According to your theory, Sir Isaac, yes—because, in your view, the gravitational field is linked to the mass. You must admit it is perfectly odd, that the gravitational field vanishes in the instant of annihilation. Don't you think?

NEWTON: Very strange, indeed. Honestly, I'm a bit confused. I believe that something happens to the gravitational field that originates from the electron-positron system. But that it should just simply vanish . . . ? I do think something is not quite correct with my theory.

EINSTEIN: There you go, Newton. With the example of the annihilation into radiation, we have really touched the Archilles' heel of your theory. We will later see that things are not quite so dramatic in my theory.

NEWTON: You're keeping the suspense up, dear colleague. I admit that my theory of gravitation cannot function as it should when we come to phenomena as exotic as the annihilation of matter and antimatter. I had no way of knowing about this phenomenon when I wrote the *Principia*. Presumably, it is too early to end our discussion of this matter right now. But I'll keep it in mind—we most certainly have to come back to this point later.

EINSTEIN: And how! You'll be surprised how simple it becomes in my theory.

Haller looked down at his watch: "Gentlemen, lunchtime is approaching. I believe that, for the time being, we have touched on enough problems. I suggest we continue our discussion after lunch."

A short time later, the three physicists were seen leaving Einstein's house and walking toward the lake, and then entering a nearby fish restaurant.

CHAPTER 4

Particles and Their Masses

If I have learned one thing from all the pondering that has accompanied me through my long life, it is this: we are further removed from a deep insight into the elementary processes in our world than most of our contemporaries would believe.

—Albert Einstein[1]

The three physicists sat in the restaurant with a glass of Berliner Weisse—a cloudy and slightly sour-tasting beer which Einstein had ordered. Newton did not relish the brew and waited for the food. He brought the discussion back to the previous subject:

"It is remarkable—mass causes itself to be surrounded by a gravitational field; an electric charge similarly generates an electrical field. But what is charge, what is mass? Haller, you mentioned before that charges appear in Nature in given units—multiples of the elementary electric charge. Some particles have a charge, others don't—like the neutron. Atomic nuclei can have large charges. The nucleus of the uranium atom has the charge number +92. We can assign a charge to every object. Very well—but in my view, a job for a bookkeeper rather than for a scientist. This does not explain what exactly is the nature of a charge; worse, it doesn't explain why charges always appear in these odd elementary units. I ask you: what is charge? And you answer that electrical charge comes in units, and could be one, two, or even zero. But that is not physics yet!"

HALLER: Hmm, I'll pretend I did not hear that remark about the bookkeeper. But in one respect you are right. Our present understanding of electrical charge is not satisfactory. We'll get back to this later. Maybe by then I can offer an explanation that will hold up to your criticism.

EINSTEIN: I also believe that this counting of charges without gaining an understanding of their nature cannot be the whole truth;

but for the time being I can live with it. After all, we can't understand everything at once.

NEWTON: So let's leave it there and turn to mass. The electron has a mass of 0.511 MeV; the proton is 1,836 times heavier than the electron. Therefore, the gravitational field around a proton is 1,836 times stronger than that which surrounds an electron. Keep this ratio in mind—and recall that the electrical fields of a proton and of an electron have precisely the same strengths. Are there particles that possess an even larger mass than the proton—I mean actual particles, not composites like the atomic nuclei?

HALLER: There certainly are—but that can't be explained in just two sentences.

EINSTEIN: It doesn't matter now. Let's proceed. What are the heaviest elementary particles that have been found so far?

HALLER: Your question takes us back to the year 1896. In that year, as you know, the French physicist **Henri Becquerel** found out that the atomic nucleus of uranium is not stable; in the course of time, it will decay, creating other, smaller atomic nuclei in the process.

EINSTEIN: Aha, that is **radioactivity**. But what does that have to do with the mass of particles?

HALLER: In the 1950s, physicists realized that the weak interaction in the atomic nuclei, which is responsible for radioactivity, is a new fundamental force of nature. It is mediated by heavy particles with lives too short to be seen directly. Some physicists assumed that this new force has something to do with the electrical interaction. Since the observed manifestations of this weak nuclear interaction is, in fact, much weaker than the electrical forces, people assumed that the carriers of this weak interaction must be very heavy. That, it turns out, will explain why the ensuing effects are very small.

EINSTEIN: All right, that means these new particles which transmit the weak nuclear force have a function just like my photons, which carry the electromagnetic force.

NEWTON: What do you mean? Are you saying that these new heavy particles would also have something to do with the photon, which, however, has no mass at all? That would be an odd relative.

HALLER: It would not be all that odd if we could find a reason why photons are massless, while the other force carriers are not. But that's exactly what people now think.

EINSTEIN: Don't keep us waiting, Haller. Tell us, what exactly are these new particles?

At this moment, the waiter appeared with the meal. The discussion was interrupted for a while. But soon enough, it started afresh.

HALLER: I'll be brief: I'll mention only those aspects that are important for what we are talking about right now. In 1973 our colleagues at CERN observed several conspicuous properties of these weak forces; they suggested that one could explain them as indirect manifestations of exactly three particles. They were called W+, W−, and Z. The **W particles** form a particle-antiparticle pair. The W− is the antiparticle to the W+, and vice versa. As the indices + and − say, the W objects are electrically charged. The **Z particle** is electrically neutral, just like the photon.

EINSTEIN: Can we say that the Z particle is sort of a "heavier brother" of the photon?

HALLER: Absolutely. It transmits a force from one particle to the next, which is similar to the electrical force. But it is much weaker, and this is a consequence of its large mass.

NEWTON: How can the mass of a particle have anything to do with the strength of the force it transmits?

HALLER: Let's assume for a moment that the photon has mass. According to the laws of quantum physics, which I do not want to cover just now, the electrical force, at very small distances, behaves exactly like in what I will call the normal case, where the photon has no mass. The electrical force is produced by the exchange of photons between electrically charged objects; these go back and forth between the charged particles and transmit a force this way. But when the distance between the particles gets large, the mass, if any, becomes important. If there is mass, the particle exchange becomes more difficult, and that means the transmitting force is weakened quickly.

NEWTON: I understand—the fact that electromagnetic forces can be effective over great distances is a consequence of the masslessness of the photons.

HALLER: Exactly. If photons did have mass, even if only minute, our world would look quite different. At any rate, no manifestations of the electromagnetic force would exist over great distances. That means there would be no such thing as an electric motor, and certainly no radio transmitter.

EINSTEIN: From what you say, I take it that the forces being transmitted by the W and Z particles—that is, the weak nuclear forces of nature, are basically about the same strength as the electrical forces, but only at very short distances. At great distances, these weaken very quickly. That is the reason why those "weak interactions" are actually weak.

HALLER: That's the idea. By observing the strength of that weak force, we can draw conclusions about the masses of the particles that transmit it. This mass must be pretty large indeed, on the order of 100 GeV. That is 100 giga electron volts, or GeV—about 100 times as much as the mass of the proton.

NEWTON: Oh really? That would be about 200,000 times the mass of an electron—really, a gigantic mass! And you believe these particle monsters really exist?

HALLER: To test if the W and Z objects really exist, a special accelerator was built at CERN around 1980. This accelerator was able to produce these particles by colliding protons and antiprotons at high energies.

NEWTON: That means that, as a result of Einstein's relation $E = mc^2$, the energy of the colliding particles actually makes up the mass of one of these new heavy particles?

HALLER: Not exactly—only part of the energy was changed into mass this way. When the experiment was done, the W and the Z particles were actually observed within a short time of each other. The W particle had a mass of about 80 GeV, the Z particle a mass of about 90 GeV. Today, we know those masses much more precisely: the W mass is 80.2 GeV; that of the Z particle is 91.2 GeV.

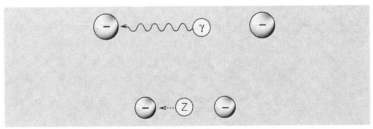

FIGURE 4.1 Forces are mediated by carrier particles: the mutual repulsion of two electrons occurs by means of the exchange of a photon, or gamma quantum. Since photons have no mass, the electric force acts over great distances. The weak nuclear force transmitted by the Z boson, on the other hand, acts only over short distances, due to its heavy mass.

EINSTEIN: Let me try and digest that. It sounds hardly imaginable—particles of a mass about a hundred times that of the proton!

HALLER: Of course, these particles do not exist as stable objects. They are produced during collisions; immediately afterwards, they decay. In order to study the decay of the Z particle more closely, another accelerator was built at CERN in 1990. This one is able to produce large amounts of these Z particles by the collision of electrons and positrons.

NEWTON: An electron and a positron collide head on and, in the process, produce a Z particle? If I use Einstein's equation, both the electron and the positron will need an energy amounting to one half of the Z particle's mass, i.e., 45.6 GeV.

HALLER: That is correct. The **LEP** machine—which stands for **Large Electron Positron** collider—is a large ring-shaped accelerator. It can hurl electrons against positrons at very high energies. In the LEP tunnel, deep underground, the electrons and positrons race, both almost at the speed of light, head-on against each other. Their orbits are calculated to lead to such collisions inside experimental halls that surround the accelerator in given locations. Matter and antimatter will then annihilate one another in these halls and, as the phoenix rises from the ashes, a Z particle will emerge. In this manner, our colleagues have produced millions of Z particles.

EINSTEIN: Wow! Electrons and positrons collide at a minuscule point in space, and out pops a Z particle with that gigantic mass!

FIGURE 4.2 The accelerator LEP on the outskirts of Geneva, Switzerland. The ring-shaped accelerator is located in a subterranean tunnel. (*Photo:* CERN)

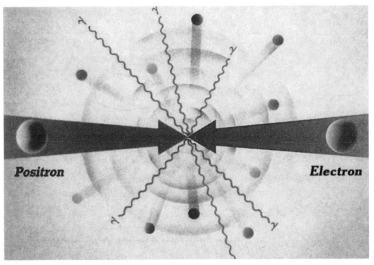

FIGURE 4.3 Schematic sketch of a particle collision at LEP. An electron and its antiparticle, a positron, collide frontally and annihilate. In this reaction, a Z particle may be produced, but it will decay very rapidly. The γ represents a gamma particle. (*Drawing:* CERN)

What an incredible energy density that must be! A regular cosmic inferno—if only in a tiny space.

HALLER: Right you are! Never before have such energy densities been produced as in these LEP collisions. In a space about the size of one thousandth of an atomic nucleus, our colleagues produce the conditions that must have prevailed at the creation of the universe—just after the Big Bang.

NEWTON: Enough now, Haller. You can sit there with a straight face and claim that the world began with a bang, sort of like the little cosmic bang that produces the Z particle at CERN? You can't really be serious? After all, the world has existed, like space and time, from all eternity. Read my *Principia*!

HALLER: Pardon me, Sir Isaac, I understand your objection. But as we will later see, the possibility of the origin of our world from one gigantic cosmic explosion is something we should take as a serious possibility. This is one of the consequences we can derive from Einstein's theory of gravitation, as we'll show later on.

EINSTEIN: This discussion is going too far. My theory is a theory of gravitation, not a theory of creation. If my theory can add to that subject, it does so only in passing, and then only with a bit of a question

FIGURE 4.4 The LEP tunnel, which has a circumference of 27 kilometers. (*Photo:* CERN)

FIGURE 4.5 The Aleph detector at CERN is one of four large particle detectors that observe the Z particle's decays at LEP. *Second from left*: Jack Steinberger, physics Nobel laureate in 1988. (*Photo:* CERN)

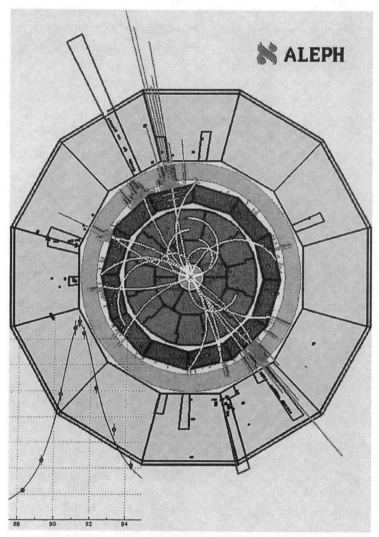

FIGURE 4.6 Computer-generated image of a cross-section through the detector Aleph. The shown particle tracks originate at the center of the detector, where the Z particle decays into several particles. The sum of their energies is equal to the mass of the Z particle. (Courtesy of CERN)

mark attached to it. I suggest we take a break to have some dessert.
Let's have no further word about cosmic physics before we get home.

During a long walk through the nearby forest, Newton followed
Einstein and Haller, lost in thought. Einstein was telling anecdotes
from his days in Berlin—an inexhaustible source of material, it
appeared. About three o'clock in the afternoon they made their way
back to Einstein's house in Caputh. His housekeeper was expecting
them, having prepared everything for tea.

NEWTON: Let's get back to mass now. We said that these monstrous Z particles pop out from electron-positron collisions in the
LEP machine. So far, so good—but what happens next? How long
do they survive?

HALLER: The Z particle doesn't last very long in this world; in
fact, it decays almost immediately.

EINSTEIN: In principle, this odd object could decay into the same
kind of particles that created it—that is, into an electron and a
positron.

HALLER: That is exactly what happens. Sometimes the Z particle
does indeed decay into an electron and a positron; but that happens
only sometimes, as I said—and it is not the most frequent decay.
There are other decay channels, such as a **muon** and its antiparticle.

NEWTON: Aha—the muons, were they not these strange elementary particles that are also unstable? Are we not using them to
demonstrate *time dilation*—one of the consequences of Einstein's
special theory of relativity?

HALLER: Exactly. Muons are the heavy brothers of electrons. As
far as Z particles are concerned, muons appear as a result of their
decays as often as electrons. The fact that muons are 200 times heavier than electrons doesn't matter in this context. The Z particle can
also disintegrate still differently, for example into protons and their
antiparticles, along with a series of **mesons** like the so-called pi-
mesons (π-mesons). In such decays, dozens of particles may appear.

EINSTEIN: How quickly, then, does one of these decays occur?
What is the lifetime of one of these Z particles?

HALLER: We can't speak of longevity. The Z particle doesn't even
stick around as long as a muon—the lifetime of which is one millionth of a second. The Z particle lives just as long as it takes a flash
of light to traverse an atomic nucleus; and that is a time as short as a
millionth of a billionth of a billionth of a second.

NEWTON: And that is what they really call a particle these days? I would say this Z particle is no more than the phantom of a particle; in fact, its pretty close to nothing at all.

HALLER: I have to say that I see that a bit differently. Times that short are nothing unusual in modern particle physics. It is even rather easy to measure such lifetimes—but we have to do it indirectly.

EINSTEIN: All right then—let's leave that where it is. Let's get back to gravity—the Z particle, after all, appears to be very heavy. That means that at the very moment it is being produced at CERN, gravitation will also come into the game. A gravitational field will build up around the Z particle. But of course, that field will be as short-lived as the Z particle itself. It will collapse when the Z decays.

NEWTON: A gravitational field is being created only to disappear again immediately—what an odd phenomenon! That would not happen in my theory of gravitation, where the field is either present or absent. Tertium non datur (there is no other way).

HALLER: No wonder. In your teaching of mechanics we saw often enough that phenomena will propagate miraculously all across space. Einstein's important realization was that it doesn't happen faster than the speed of light. Every signal in the universe, be it lightning or whatever, propagates with a velocity that cannot exceed the speed of light. Which, to reiterate, amounts to 300,000 km per second, or 186,000 miles per second. There is no faster velocity. Now, for the sake of argument, let's assume somebody could make the Moon disappear all of a sudden.

NEWTON: Okay, so the Moon disappears, and its gravitational field vanishes with it—or do I understand you to mean that the field doesn't vanish without further ado?

HALLER: Surely not. It takes light one second to travel from the Moon to us. It is not difficult to measure the Moon's gravitational effect here on Earth. Let's assume the Moon vanishes precisely at midnight tonight.

NEWTON: I'm getting it: at midnight precisely the gravitational field of the Moon still exists on Earth; the same is true immediately afterwards, and only one second after midnight will it have disappeared.

HALLER: The gravitational field of the Moon won't be able to vanish suddenly from all of space. It will disintegrate continuously, but with the velocity of light. For a person like Newton, who is after all the discoverer of the inertia principle in mechanics, this will not appear strange. This disintegration occurs in the form of a gravitational shockwave. This shockwave propagates spherically from the original position of the Moon, with the velocity of light. It is similar to a wave that we create when we throw a stone into a pond. This shockwave reaches the Sun after eight minutes, and will take as long as five hours to travel as far as the planet Pluto, in the outer reaches of our solar system.

EINSTEIN: It was clear to me, when I developed my special theory of relativity, that the gravitational field does not behave in a way that is qualitatively different from the electric field. A field is just that, a field. The word means that the action of gravity which we observe at any given point is, in fact, a property of the very space (or region of space) where we observe it. The Sun attracts the Earth because the Sun changes the space around the planet Earth. The Sun's gravitation builds up a field of force, but it is not the Sun that acts on planet Earth over this huge distance—no, the Sun cannot do that.

The Earth moves in the gravitational field of the Sun, but this field has its own properties. Primarily, the field is a property of the space in which the Earth moves. Never mind that the field is created by the Sun. It cannot suddenly be turned off—that can be done only over some time lapse—the time lapse needed by a light signal to propagate across the relevant spatial distance. The same is true when we talk about the Z particle. At the instance of its creation, a gravitational field is being built up; very shortly thereafter, as the Z decays, that field will also collapse.

NEWTON: This gravitational effect doesn't get very far—with great difficulty, it reaches the size of an atomic nucleus.

EINSTEIN: That doesn't matter. Let's just concentrate on the principle. After all, we can't have the Moon vanish, or be re-created, all of a sudden. For a Z particle, however, that can be done. When I developed my theory here in Berlin, I had no way of knowing about any of these remarkable elementary particles. Our discussion about the Z particle made me wonder about what mass actually does mean. Let's face it, here we have a really heavy object, a true monster of mass—indeed, of pure mass, I might say—about 100 times heavier than a proton. Isn't this our chance? Couldn't we take a closer look at this Z thing just to find out what exactly mass is? Mass, we already know, is the source of gravitation. We do know what effects mass has, but not what mass actually is. Whence does it come? How does it equip the Z particle with this huge value?

HALLER: Let me admit one thing right away: as much as I would love to give you a straight answer, I cannot do that. To this day, we don't know what mass really is. The whole thing is more mysterious if you bear in mind that the Z particle isn't even the heaviest elementary object we have discovered so far. As mentioned before, the atomic nucleus is not an elementary (i.e., a structureless) particle. Rather, it consists of even smaller structures, which we call quarks. However, the latter can't be observed as free particles because the forces between the quarks are so strong that they cannot be isolated from the other quarks. But we can observe them without difficulty inside the nucleus. Quarks, however, do have mass. In this respect, they behave like other particles.

The quarks that make up normal nuclear matter have relatively small masses—small enough to permit us to neglect this aspect for a number of purposes. But there are other quarks, more exotic, with much larger mass. The heaviest quark, the t-quark, was discovered in 1994 from collisions of protons and antiprotons at the **Fermi National Accelerator Laboratory** near Chicago. There, it is possible to observe head-on collisions of protons and antiprotons with an energy of 1,000 GeV each. Occasionally, if not very often, such collisions will result in the creation of one of these t-quarks and its antiparticle—the t-antiquark.

NEWTON: And how large is the mass of this t-quark?

HALLER: As I said before, this is the heaviest object we have found so far. Its mass is about 180 GeV, almost twice the Z mass. It is really an enormous mass, and physicists are duly puzzled. Even prior to their experimental detection, it was possible to estimate the masses of the Z and W bosons. For the t-quark, that was not possible. We knew there must be a t-quark before we had ever seen it. But all estimates of its mass were in the 15 to 50 GeV region—far from the final result.

EINSTEIN: I assume that the t-quark is just as unstable as the Z boson. So it must decay into other particles right after its creation. True?

HALLER: Its lifetime *is* briefer, but not much briefer, than that of the Z and W bosons.

EINSTEIN: Isn't it amazing that the mass of the t-quark is about twice that of the Z particle? Maybe we can take this as a hint that

FIGURE 4.7 Aerial view of the Fermi National Accelerator Laboratory, west of Chicago. The large ring is what can be seen from above of the accelerator, presently the world's highest-energy particle accelerator. (*Photo:* FNAL)

FIGURE 4.8 The detector where the t-quark was discovered. (*Photo:* FNAL)

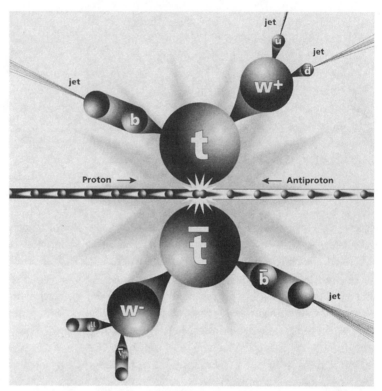

FIGURE 4.9 Schematic view of the production of a t-quark and its anti–t-quark particle in a proton-antiproton collision. The t-quark decays immediately after its production into a lighter quark called the b-quark, plus the W particle, carrier of the weak force.

there must be more connections. This system of superheavy particles, like the W, Z, and t particles, could be the Gordian knot of the mass problem. So we will just have to learn to undo it. But how do we go about it? Does anybody have an idea? Not that I need to know it precisely. I'll be happy to get a few relevant hints.

HALLER: You are not letting me get away easily. As I said—to date, we have no clear notion. I would also be happy if we had an inkling of why the ratio of masses between the Z and the t particles is about 1 to 2. Nevertheless, we believe that today's research is heading in the right direction for a solution of the mass problem. But it is not an easy task. It cannot be done without accelerators—which, unfortunately, cost a great deal of money.

An interesting idea that might help to solve the problem of mass is the hypothesis that mass has something to do with the structure of the vacuum. An electric field influences the vacuum—i.e., empty space—such that in every point of that space an electric force can act. The strength of this force, which we call electric field strength, can easily be measured. In the same way, imagine that the mass of a particle—say, the electron's mass—also reflects a property of the vacuum which is described in terms of a special field.

NEWTON: Would that mean that otherwise empty space is, so to speak, filled with a field which has only one task: to tell the electron, or to tell a t-quark, what mass they are supposed to have?

HALLER: You might say so. This field would be responsible not just for the electron mass but also for the masses of the Z, the W, and all the other elementary particles. But that is a long story. Let me make a suggestion: one purpose of my presence in Berlin is a lecture that I am supposed to give at Humboldt University. In that lecture, I will address the mass problem. Since it is scheduled for this evening, let me suggest that you accompany me. We can continue our discussion tomorrow.

In the late afternoon, they took the light rail from Potsdam to Zoo Station. It was a lovely summer evening. Thousands of people were walking up and down Kurfürstendam. Einstein obviously enjoyed being back in Berlin, where he had done some of his most successful scientific work. Our three physicists walked along this wide artery of Berlin life. They had to meander sideways to reach Wittelsbacher-strasse. In the house at Number 13, Einstein had had a small apartment in the beginning of his Berlin period. During World War II, the house had been destroyed. Now Einstein stood there silently for several minutes and tried to recall details of that period.

Haller called him back to the present when he said: "So this is the place where the general theory of relativity saw the light of this world in 1915. This is, to a certain extent, the general relativity analogous to Kramgasse 49 in Bern, Switzerland."

Newton noted, not without irony: "Had this house not been destroyed in the war, I am sure we would now be standing in front of a commemorative plaque, just like the one in Bern."

Einstein replied: "You are right. But far as the commemorative plaques are concerned, I guess I was spared. As you know, the creation of my theory was a delivery by forceps; it dragged on for years, and there had been some miscarriages. There was no such thing as a sud-

FIGURE 4.10 Entrance to Humboldt University in Berlin where Einstein taught during his stay in the German capital. (*Photo:* Humboldt University)

den flash of insight. But let's leave this place. Not much remains here that would still attract me. I propose we have a little snack at the Café Kranzler before we go to Haller's lecture.

An hour later, Einstein was seen with his two friends, walking briskly across Tiergarten Park, in the direction of the Brandenburg Gate. At Paris Square, they reached the splendid old city axis called "Unter den Linden"; and a few minutes later, they entered the front gates of Humboldt University. This is where Einstein had lectured during his Berlin period. These lectures of his were one of the great attractions of the metropolis in the 1920s. The first lectures on the theory of gravitation had also been given at Humboldt University. Einstein knew his way around the large building and had no trouble locating the lecture hall one floor up—where Haller's lecture was to be presented.

CHAPTER 5

Haller's Lecture: Empty Space and Modern Physics

The most beautiful experience we can have is based
on mystery. That is the sensation from which both
art and science originate.

—Albert Einstein[1]

"Ladies and gentlemen!

"Imagine this is the year 1654. The Imperial Diet is meeting in the
city of Regensburg; there are representatives of German princes, and
those of the professional classes. From Magdeburg there is **Otto von
Guericke** (1602–1686), *Bürgermeister* (mayor) of his home city. But
Guericke is not only the mayor of the city, he is also an excellent natural scientist—he is one of the first to study the phenomenon of air.
Textbooks on technology list him as the inventor of the airpump.

"In front of visitors at the Regensburg Imperial Diet, Guericke
carries out an experiment that is about to become a hallmark in the
history of the natural sciences and technology. He joins two hollow
hemispheres that fit precisely at the perimeter; he sucks the air out
of the joined volume of what now becomes an empty, hollow sphere.
The two hemispheres, which would easily be separated under normal conditions, are now, it turns out, tightly pressed to each other by
the external air pressure; they can no longer be separated by the
physical force applied by one person. It takes eight horses to pull
them apart!

"Guericke was the first to create a 'vacuum' on Earth—although
today, it would not qualify as a very good vacuum. The word *vacuum*
stands for empty space, space without air molecules inside. To date,
nobody has been able to produce an ideal vacuum. The best vacuum
pumps available will reduce the density of air molecules to a minuscule fraction of the amount usually prevailing. But even when we

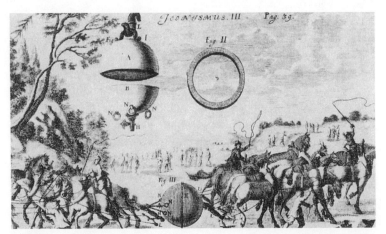

FIGURE 5.1 The experiment with the Magdeburg hemispheres was executed by Otto von Guericke at the Imperial Diet in the city of Regensburg in 1654. (Contemporary drawing

speak of an ultra-high vacuum, we are still dealing with about 100 million molecules per cubic centimeter.

"To experience an almost ideal vacuum—say, a volume of space that has fewer than one molecule per cubic centimeter—we would have to move very far from Earth into intergalactic space. Today, however, we know that even those faraway regions of space are not totally empty; there may not be gas molecules, but that space contains many photons—'particles of light.'

"In contrast to particles of matter, say, protons or electrons, photons are massless. This permits them always to move with the velocity of light across space, transporting energy. In the universe, we find, in the mean, about 400 photons per cubic centimeter. Compared with the density of air molecules at the surface of the Earth, this is very little—but it is a large density when compared with atoms in the universe. If we were to distribute evenly all the atoms of which the planets and the stars are made, across the universe, there would be only about one atom per cubic meter. Photons are the particles most frequently found across the universe. There are about ten billion more photons than atoms.

"Modern cosmology has a simple answer to that. Photon radiation, which fills all space evenly, is really what remains of the Big Bang. That was the explosion, or event, which occurred approximately 14 billion years ago. Our world as we know it today origi-

FIGURE 5.2 The Andromeda galaxy in the constellation of the same name. About 2 million light years from Earth, this galaxy of more than 100 billion stars is the one that most closely neighbors our own galaxy. The intergalactic space between Earth and Andromeda galaxy is filled with photons having the wavelength of radio waves.

nated at that time. Shortly after the Big Bang, matter was very hot; a large fraction of the energy density of the universe consisted of electromagnetic radiation, or the radiation of photons. This photon radiation also 'expanded,' cooling down in the process. Today, its temperature does not exceed 2.7 degrees Kelvin—that is, 2.7 degrees above **absolute zero.**

"If we were to examine more closely a volume of space outside our star system or our galaxy, we would discover that, strictly speaking, we're not dealing with a vacuum at all but rather with space that is filled with photon gas. In principle, it should be possible to extract all the photons. The space we are considering would have to be surrounded by a metal coating that prevents photons from entering. But in order for that to work, we would have to cool down the metal coating to absolute zero; that is possible in theory, impossible in practice. And even then we would not have an absolute vacuum. Modern cosmology tells us that, on average, all space has not only an energy density of 400 photons per cubic centimeter but also some 500 **neutrinos.** Neutrinos are neutral particles that are related to the electrons that make up the shells of atoms, but do not have an electric charge.

"We know three different kinds of neutrinos. The first is a partner of the electron in the weak interaction, the same interaction that is responsible for radioactivity. It is therefore called the *electron-neutrino.* The second neutrino is a partner to the muon, and we call it the *muon-neutrino.* The third one is the neutral partner of an additional brother of the electron, which we call the *tau-lepton* (τ-lepton), with a mass that is almost 20 times that of the muon.

"Neutrinos possess either no mass or, at best, a mass that is only a tiny percentage of the mass of the electron. To date, we have not been able to prove that neutrinos have mass, although there are some indications. Until proven otherwise, we will consider them as massless.

"Because of their lack of electrical charge, neutrinos interact very rarely with matter. A neutrino can, without any effort, travel through the entire Earth without interacting with a single atom or molecule. Photons can be shielded, neutrinos cannot. This means that the evacuated volume of space we mentioned earlier may contain neither atoms nor photons, but it will certainly contain neutrinos. Neutrinos are constantly passing through.

"In principle, it is impossible to cut this ever-present neutrino flux. And that means there is no way to have a true vacuum—a truly empty space—on Earth or anywhere else in our universe. We can only imagine it as a theoretical construct—an abstract concept we can never realize. Still, we can imagine it, and base useful considerations on this idea.

"Modern physics describes matter in terms of particles that move in empty space. Space, if you wish, is like a container in which matter particles, such as electrons, are embedded. When there are ten electrons in a volume of, say, one cubic centimeter, it will be easy to add another electron, or to remove one. This will not change the space as such. Therefore, in principle, we can imagine that we can create a complete vacuum, a really empty space, by removing one by one all the particles that are initially in that volume: electrons, photons, neutrinos. The result of this theoretical construct would be what we call 'pure space'—the ideal vacuum. It is the vacuum that resembles the idea of empty space discussed more than two thousand years ago by Greek mathematicians, such as Euclid: our space simple and pure with its three dimensions.

"There is one thing we forgot when imagining how to create our ideal vacuum. Physics has not only particles, it also has physical fields. We know that we can influence the motion of an electron with a magnetic field, which we can generate with a coil that carries an electric current. The magnetic field then fills space—more precisely, it becomes a physical property of that space. In order to generate the ideal vacuum, we have to be sure that the observed space is not only free of particles but also free of fields. This can be reached, in principle, by suitable shielding. The ideal vacuum then is a volume of space that has neither particles nor fields.

"We might now conclude that this vacuum, which we built up with such great care, is ultimately a mathematical construct that has nothing but these three dimensions, but no physical properties. This idea was shared by physicists up until the 1930s. All this changed rather quickly with the development of quantum theory. More precisely, it is the unification of quantum theory and Einstein's theory of the relativity of space and time that changed all this. It turned out that the interpretation of the vacuum as a passive, empty space—as a container in which we can simply store particles of matter—is just not tenable. This reorientation of how we should look at the vacuum is closely linked with the name of **Paul Dirac**. He did his research in Cambridge as early as the 1930s. Later on, he occupied

FIGURE 5.3 Paul Dirac, whose theories describe the properties of "empty" space. (*Photo:* Cambridge University)

the chair of physics in this university town that had once been Sir Isaac Newton's.

"Dirac tried, in 1928 or thereabout, to combine **quantum mechanics**—which had been developed by **Werner Heisenberg** and Erwin Schrödinger and had given the first precise description of atomic processes—with Einstein's theory of relativity. It became clear to him that it was not an easy task to make them compatible.

"The typical velocities of the particles inside atoms are a good deal smaller than the speed of light, which as we all know is about 300,000 km per second, or 186,000 miles per second. For that reason, the effects of relativity theory—which we call relativistic phenomena—have little or no importance in atomic physics. Dirac did not set out to develop a new way of looking at the physics of atoms; rather, to him it was a matter of principle. He wanted to generalize the newly developed quantum theory to processes where the velocities of the particles involved were to come close to the speed of light. Of course, he would not imagine that, just fifty years later, physicists would be able to accelerate particles to velocities which are very close to the speed of light.

"An important characteristic of Einstein's theory of relativity is the lack of any essential difference between space and time—whereas in classical physics, they are completely distinct. In a certain way, space and time can mix. In relativity theory, there is no neat distinction between space and time. They make up one single phenomenon, the spacetime of Einstein.

"In 1928, Dirac managed to derive a mathematical equation that provided the essential link between quantum theory and the theory of relativity. Dirac's first result with this equation was crowned by an impressive success: he was able to derive with great precision the strength of the interaction between the electrons in the atomic shell with magnetic fields. His equation made it clear that here was an important step in the direction of a deeper understanding of elementary particles. It has been called 'Dirac's equation' ever since.

"Soon after, he noticed that his equation contained in its solution not only the properties of electrons inside atoms; since his equation, just like relativity theory, treated space and time as equal partners, he found solutions to the equation for positive and negative energies. The positive energies are to be assigned to the well-known electrons. The negative-energy solutions caused him considerable headaches. In quantum theory, it is quite common that a given atomic state makes a transition to another, lower-energy state as long as the relevant conservation laws permit it.

"The light from a television screen which remains visible in a dark room after the set has been turned off is caused by the same phenomenon. When the TV set is on, the electron beams that generate the image on the screen excite the atoms in the screen and raise them to a higher-energy atomic state. But when the set is switched off, these atoms will revert fairly quickly to their state of lowest energy, in the process emitting no-longer-needed energy in the form of visible light. As we can easily see, this phenomenon keeps going for several minutes.

"An electron at rest described by Dirac's equation has an energy of $E = mc^2$, corresponding to Einstein's famous relation between mass and energy. In Einstein's theory, the energy of a particle is always positive. This is not true in Dirac's theory, which, as we said before, allows objects with negative energy. Even more so: for every particle in Dirac's equation, there is another particle which has precisely the same energy but now with a negative sign. Since mass and energy are equivalent, we might just as well speak of a negative mass.

"Initially, Dirac tried just simply to ignore the negative-energy particles his solution suggested. But soon he realized that that was impossible. According to his equation, an electron at rest could change into an electron with negative energy $-mc^2$, by radiating off the energy difference of $2mc^2$. Note that in this process, the total energy is conserved. The initial energy is mc^2; the final energy is the same, $mc^2 = 2mc^2 - mc^2$. We can even calculate how quickly such a conversion process will occur. It would take only about a hundredth of a millionth of a second. Of course, this is pure nonsense since we know that a free electron at rest does not engage in such follies—it remains in its initial state as long as it likes.

"Dirac was in a dilemma. On the one hand, his equation had great success in atomic physics; on the other, it resulted in absurd descriptions of electrons in the vacuum. There were only two possibilities: either the equation was wrong, or another interpretation of the results of the equation was needed. To pull himself out of this quandary, Dirac decided to try the latter. The result was a completely new interpretation of the vacuum, of empty space.

"The behavior of the electron inside an atom can only be understood when we assume that two electrons are never at any given place at the same time. This important principle of atomic physics, dubbed the *exclusion principle*, was introduced by **Wolfgang Pauli** in the mid-1920s, and it has stood the test of time. It is also well known as the **Pauli principle**.

"If we picture electrons as tiny metal spheres, we can easily interpret the Pauli principle: wherever there is a speck of matter, there is no room for any other matter at the same time. In atomic physics, as described in terms of quantum theory, this is not necessarily obvious. In quantum physics it is not possible to say with absolute certainty where exactly a particle is located. We can only say that with a given probability—say, with a 50 percent likelihood—it can be located in a certain place in space. This is a consequence of Heisenberg's **uncertainty principle**. It says that, in quantum mechanics, we cannot make any absolute, precise statement concerning the location or **momentum** (velocity) of a given subatomic particle.

"For that reason, it would appear that any two electrons could have the same chance or opportunity to exist simultaneously in one and the same location. For some particles, like the photon—the particle of light—this is indeed the case. But for the electron, this possibility is excluded by the Pauli principle. More precisely, that principle says that two electrons *cannot* be in the same quantum state. Wherever there is one electron, there cannot be another one.

"Strictly speaking, the Pauli principle is the basis for the fact that two pieces of matter—for example, a piece of wood and a piece of iron—cannot coexist in the same place. Actually, this might be possible because matter, built up from atoms, is not that densely packed. Two atoms come close to each other in a rigid body; but the space inside the atoms is basically empty: atomic nuclei are a good deal smaller than the atoms that include the electron shell. But the reciprocal penetration of individual atoms is prevented by the Pauli principle. When we knock our heads against the wall and notice this by the pain, the deeper cause is not a repulsive physical force between head and wall: rather, it is the Pauli principle.

"Dirac discovered that his equation is acceptable if one assumes that the vacuum gives up its passive role, which it has in classical physics. Since, according to the Pauli principle, no electron can occupy a location already taken by another one, he assumed that the negative-energy states suggested to him by his equation for some of the electrons do actually exist. A positive-energy electron might seek to change into a negative-energy state by emitting electromagnetic radiation, but it is unable to do so because the available positions in the space that it covets are all taken.

"This becomes possible only if we give the vacuum a different meaning: *it is the quantum physical location of all permitted electron states with negative energies—it is the quantum state of the vacuum.* In this connection, we also speak of the vacuum in terms of a 'sea' of negative-energy states. If, however, we notice an individual electron—say, an electron at rest—that has an energy corresponding to its rest mass of about 0.5 MeV, Dirac will tell us that this state reveals more than what concerns that particular electron. Rather, it indicates a state where all negative-energy locations for electrons are taken, and one positive-energy electron state is also occupied.

"We can see right away what Dirac's new interpretation of the vacuum means. The vacuum relinquishes its soul as 'nothing' and turns out to have a number of interesting physical properties: the vacuum has its own peculiar existence. Let's assume, for example, that we add to this vacuum state a certain energy—an energy of about 1 MeV—that is the energy corresponding to two electron masses. It can now happen that one out of the infinity of electrons in the 'Dirac sea' will pick up that single MeV and add it to its own negative energy of -0.5 MeV. This means it gets 'excited' and changes into an electron with a positive energy of +0.5 MeV. This would be an electron at rest. But the state we have now reached is not exhausted with the description of a single electron. Recall, we have pulled an electron from the vacuum by adding energy. That electron is now missing from the vacuum: it leaves a 'hole' there. We now no longer have a normal vacuum but rather a state with a missing electron. This situation is analogous to a bank account. The state with the electron corresponds to the account with a balance of, say, 100 dollars. The actual vacuum corresponds to the account with a balance of zero; the vacuum with a hole might stand for an overdrawn account having a 100 dollar negative balance.

"When we compare this 'vacuum with a hole' with the normal vacuum, it might be seen as an entity in which the electric charge has been lowered by one unit of charge. At the same time, its energy is greater than that of the normal vacuum, by precisely 0.5 MeV. This odd structure looks almost like a particle, but a particle that has a charge, that is equal to that of an electron, but with a positive sign. This is what we call a positron, the antiparticle of an electron. We should not be disturbed that this state 'with a hole' has a positive charge; we might identify it as a missing negative charge. The assignment of 'positive' and 'negative' was introduced in the eighteenth century by Benjamin Franklin. In his time, it was not known that electric currents had anything to do with the motion of particles that we now call electrons. Otherwise, the sign of the individual charges would have been defined differently: the electron, being the principal agent of all electrical phenomena, would have been assigned a positive charge. That would simply imply that the positron would wind up with a negative charge. Seen from our vantage point, this would have been a logical choice—but hindsight doesn't help. The assignment of a given sign for a charge is really arbitrary; we have no trouble living with Franklin's 'wrong' assignment.

"You see: we added an energy of 1 MeV; in this way, we changed the vacuum to a state that contains both an electron and the corresponding 'hole,' which is a positron. This means we produced a pair of particles—more precisely, the electron particle and its antiparticle. This process is called *pair production*.

"We expect that the state thus created, which consists of a particle and its antiparticle, can revert to the original vacuum state by

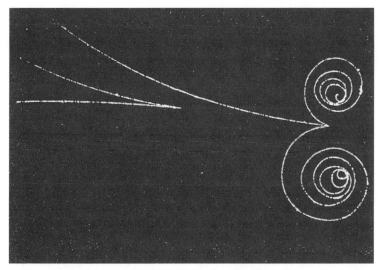

FIGURE 5.4 Production of an electron-positron pair in the vacuum. The different charges of the two particles produced can be seen from the opposite curvatures of their trajectories in the magnetic field. (*Photo:* CERN)

radiating off energy. That means there is not only the pair creation process but also one of pair annihilation. An electron and a positron that approach each other can annihilate; the energy of both particles will then be radiated off electromagnetically. You might say that, in this process, the electron jumps back into the hole of the Dirac sea. And that means: the credit side and the debit side in our account balance precisely.

"So far, we have described the consequences that Dirac drew from his new way of looking at the vacuum. His most interesting conclusion was this: the electron must have an antiparticle with a mass that is exactly the same as its own, but with a positive electrical charge. Since 1932 we have known that this particle, the positron, in fact does exist. That is the year **Carl Anderson** analyzed cosmic rays at the California Institute of Technology in Pasadena and discovered the positron in the process. In this fashion, a new door was opened in the natural sciences, the door to the world of antimatter. Today, we know there are antiparticles for all observed particles—for example, antiprotons and antineutrons.

"The processes of pair creation and pair annihilation are well known in today's physics. Whenever we manage to concentrate a sufficiently large amount of energy in a point of the vacuum, there can be spontaneous pair creation. It is not possible to predict precisely what the particles thus produced will do—say, in which direction they might go.

"It is a remarkable fact that, in their birth process, the electron and its antiparticle will appear with one precisely known mass. We can, if you wish, pull an electron-positron pair out of the vacuum by colliding two photons—that is, two particles of light. In the instant of collision, the two light quanta will give to the vacuum the necessary energy; this energy must be larger than, or at least equal to, twice the electron mass, or 1 MeV. That opens the door to the appearance of the particle pair out of 'the nothing.'

"One might ask: how can the vacuum 'know' what the masses of the electron and its antiparticle need to be, when they are thus produced out of the 'nothing'? Dirac's theory gives a simple answer. Since the vacuum is in fact an infinite sea of virtual particles, it already has all the information it needs about electrons and positrons. The full physics of these particles is contained in the vacuum itself. For Dirac's vacuum is much more than the vacuum of classical physics, say, Otto von Guericke's vacuum. The laws of physics that are important for the electron, and all its properties, such as its mass and its charge, are already present in the vacuum—not in actual reality but as virtual possibilities. *It is therefore a physical medium.* It is considerably more than the vacuum of classical physics, which is a volume without any properties, without any matter.

"Dirac's concept of the vacuum has one unattractive quality, however. Since the vacuum is described as a sea of electron states with negative energy, one would assume that the electrical charge of the vacuum is infinitely large. In physical processes, we always deal with nothing but the differences of electrical charges of the particle present, in comparison to the vacuum. This is not an immediate problem, since infinities do pop up in physics here and there; but it does appear to give a lack of symmetry between particles and antiparticles. In Dirac's picture, we speak of an electron sea and not of a positron sea. The electrons appear to get preferential treatment.

"This lack of symmetry was taken care of by further refinements of Dirac's theory in the 1930s; that is when the theory of relativity and quantum mechanics were finally fused together into what we call *quantum field theory.* The development of this theory is owed, above all, to Werner Heisenberg and Wolfgang Pauli.

"If we examine Dirac's equation in the framework of quantum field theory, it turns out that we can write down a mathematical equation that describes the positrons—the particles that pop up as holes in the Dirac sea. This equation is identical to the original Dirac equation once you replace electrons with positrons and, in the process, change the sign of the electrical charge. That means Dirac might have started his examination with this equation. But in that case, he would not have interpreted the vacuum as a sea of electrons with negative energy, but rather as a sea of positrons with negative energy. The electron would then be a hole in the sea of positrons.

"In quantum field theory, we unify both pictures; we characterize the vacuum not only as a sea of electron states of negative energy or of positron states of negative energy. Rather, we see in it a sea of (virtual) electrons *and* positrons. In this way, the problem of the electrical charge of the vacuum is immediately resolved. Since there are positively charged positrons as well as negatively charged electrons in that sea, the vacuum, on average, is electrically neutral. We now have complete symmetry between particles and antiparticles.

"The sea of electrons and positrons with negative energy in the vacuum is not just a picture that we need for a mathematically consistent description of particles. Rather, it leads to concrete physical consequences. Let's assume we'll add an electron to this vacuum. The electron has a negative electric charge. This charge acts on the electrons and positrons in the Dirac sea: it repels electrons and attracts positrons. In this fashion the added electron causes a distortion of the vacuum; in physics we say it *polarizes* the vacuum. The positrons in the vicinity of an electron will partially shield its charge, so that the charge of the electron will look smaller from a distance. We describe this phenomenon by saying: the 'bare' charge of the electron (i.e., the charge of the electron in the absence of a Dirac sea) is larger than the charge we actually measure.

"We can observe this effect when measuring the deflection of two electrons as they fly by each other at close range. When these distances become very small, the effect of the Dirac sea can be partially ignored. In fact, we find that the electric charge of the electron, if measured at distances at least 10,000 times smaller than the radius of the hydrogen atom, is a bit larger than at greater distances. The effect of this increase in electric charge can be calculated. The results of measurements of this effect confirm these theoretical predictions completely. We can therefore say: the Dirac sea is not just an abstract invention of theoretical physicists; it is a physical reality.

"In modern elementary particle physics, of course, physicists go much further than Dirac with his electron-positron vacuum. The vacuum of today's physics comprises not just electrons and positrons of negative energy; rather, *all* elementary particles are virtually present in the vacuum. It also contains quarks and antiquarks, the smallest particles of the atomic nuclei. When producing a high-energy density in the vacuum, we don't just pair-produce electrons and positrons as we discussed before. Rather, we will also produce quarks and the corresponding antiquarks. These processes can be observed at the LEP accelerator at CERN in Geneva, Switzerland.

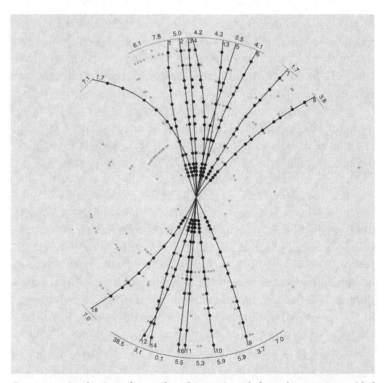

FIGURE 5.5 Production of a quark and an antiquark from the energy provided by the annihilation of an electron and a positron that had previously been accelerated by the PETRA accelerator in Hamburg, Germany. With a velocity close to that of light, the two quarks move apart, decaying into two "jets" of particles that can be observed in the particle detector. (DESY [Deutsches Elektronen Synchroton], Hamburg)

"Fundamentally, all the elementary particles that make up matter, or that we need to understand the forces of nature, are already present in the vacuum; we might say (to be more precise), they are virtually present there. Say, for example, we use energy in order to produce a muon, the heavy brother of the electron, and its antiparticle. We are not really creating it out of the 'nothing'; both particles have been present in the vacuum before, although only as virtual objects. The energy we need to make this process happen is simply the price we pay to lift the two particles out of the inexhaustible supply of the vacuum.

"Even more, its not just the particles that are present in the vacuum; the same goes for all the essential properties of our world—the laws of nature that fix the rules for the dynamics of all natural processes. Those laws that we derive from our observation of visible matter in our world are not tied to this matter. They really are, ultimately, properties of the vacuum, properties of space which seems to be completely empty.

"It took scores of years before this idea was fully accepted by the physics profession and before it was realized how important its consequences are for astrophysical and cosmological problems. We know, for instance, that an electron in some faraway galaxy has exactly the same properties as an electron here on Earth. If we could carry out experiments in that galaxy—say, pair creation of an electron and of a positron—we would be sure to find the same results as here on Earth. This would be hard to believe were it not for the fact that these physical laws are actually the properties of space itself; that is, properties of the vacuum.

"Physicists did not stop there. The idea of a Dirac vacuum filled with particles and antiparticles does not explain why elementary particles—like electrons or quarks—have mass. According to Einsteins's relation between mass and energy, $E = mc^2$, we know that the mass of a particle can, under given circumstances, be transformed into radiative energy. This happens, as we know, automatically in pair annihilation. This does not prove that a particle has to have mass. It might well be massless, like the particle of light, the photon. It is also possible that the neutrino particles don't have a mass.

"Theoretical physicists like to imagine that, next to the real world, they have to describe an ideal one in which the laws of nature are a bit less complicated than in reality. In this vein, they often consider the case where particles have no masses. Many processes are then much easier to calculate for them, since the troublesome mass of particles can be ignored. Since about 1970, we have learned that we can come to a unified description of elementary particles and their interactions when, to start with, we ignore their mass. Later on in the follow-up reading of the laws of nature, we can then add mass as small perturbations.

"This step, of course, does not come for free. We need an additional force of nature to complement electromagnetism, the strong and weak nuclear forces, and gravitation. This new force is mediated—how else?—by a particle we have not seen yet in experiment to this day. This is the so-called **Higgs particle**, named after the British physicist Peter Higgs. He is the one who 'invented,' besides others, the idea of this particle in the 1960s. If we subscribe to his hypothesis, which has not been proven to this date, the mass of any elementary particle is nothing but a manifestation of the strength with which this new 'Higgs force' acts on the particle.

"The Higgs force adds a decisive feature to the vacuum. If truly present, the field associated with the Higgs force is the only one that influences the vacuum directly. At every point in space, this field has a fixed value that will define the masses of particles. Physicists imagine a rather curious way in which this mass value comes about. We can illustrate it with an example from mechanics.

"Let's look at a little ball lying at the center of a structure that looks like a Mexican hat (see fig. 5.6). It is located in an unstable equilibrium. The slightest motion will cause it to leave its central position, leading it to roll down into the surrounding ditch. Once there, it will come to rest at some point. It will then be at rest in a state of stable equilibrium. Note that the point where it comes to rest is now removed from the origin, the center of the figure; that is, the original circular symmetry has been broken.

"What does this simple example from mechanics tell us about the Higgs field? That it is a spontaneous breaking of a symmetry: at the origin, where it started, the ball was in the very center of the system, the Mexican hat; its distance from the center of circular symmetry was zero. Now, as soon as the ball rolls away from the origin, the symmetry is broken. Once it comes to rest at some point in the ditch, there is no circular symmetry anymore. Remarkably,

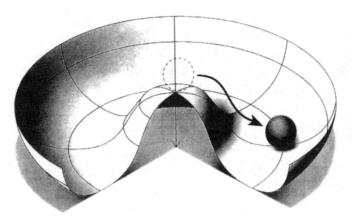

FIGURE 5.6 A mechanical model for symmetry-breaking: a sphere sitting at the center of a "Mexican hat" is in an unstable state of equilibrium. The slightest vibration causes it to roll down from the center of the hat into the surrounding ditch.

the nonsymmetrical arrangement corresponds to a stable equilibrium, unlike the symmetrical one, which is unstable.

"Let's imagine that the Higgs field in the vacuum can show a similar behavior. For once, it might be in an unstable but symmetrical configuration. In this case, all particles are massless; the world is simple and has a high degree of symmetry. This is Nirvana for physicists. But the symmetry will be broken when the Higgs field (the 'Mexican hat' field of fig. 5.6) assumes a given value in the vacuum. In our illustration, this value can be read off the figure once we know the distance between our sphere and the origin. Our particles—say, the electron—are now acquiring their masses. These masses are a consequence of symmetry-breaking.

"According to this picture, we assume that the vacuum state in our world corresponds to a nonsymmetric state of stable equilibrium. The Higgs field will have the same vacuum expectation value anywhere in the universe. This is the reason why the mass of an electron in a faraway galaxy has the same value as it has on Earth.

"We mentioned that, to this day, we don't know whether this simple idea of spontaneous symmetry-breaking really is at the bottom line of mass creation. In the near future, we will probably have the answer. The Higgs particle, the carrier of the Higgs field, might show up in experiments at today's particle accelerators. Ever since the LEP accelerator at CERN was put into action in 1989, physicists have been looking for this hypothetical particle, without success. More favorable are the chances that the Fermilab accelerator TEVATRON near Chicago or the next CERN accelerator—presently under construction—the Large Hadron Collider (LHC)—will be able to find evidence for its existence.

"The method of spontaneous symmetry-breaking for the production of masses assigns important new features to the vacuum. It is not only, as stressed by Dirac, a place where particles and antiparticles with negative energy can roam around; it is also a medium that is responsible for the breaking of symmetry, and therefore for the assignment of mass to our elementary particles.

"We might ask whether the Higgs field could not also exist in a symmetrical configuration. In our present world, this is surely not the case. But physicists expect that the symmetrical case will prevail when the energy density becomes very large. The breaking of the symmetry would, so to speak, melt away. It is impossible to demonstrate this 'melting' of the vacuum through the use of a particle accelerator: the needed energy density is way beyond what can be procured with an accelerator. Still, cosmology might provide circumstances where this process can happen.

"Billions of years ago, the matter of the universe was created in the Big Bang. Shortly after the Big Bang, the energy density was so large that symmetrical vacuum configurations formed. This has interesting consequences for cosmology. In the symmetrical configuration, the vacuum has an enormous energy density, causing the universe to expand rapidly. This process is so fast that we no longer speak of a simple expansion of the universe: rather, we speak of **inflation.** An interesting consequence of this inflation would be that all available irregularities in the universe—say, that of energy distribution—are removed. They will just be 'smeared' over huge regions. This could be the explanation for the remarkable phenomenon that our universe is quite homogeneous as far as our astronomers can observe it. We mentioned the photon radiation that fills our universe—it is perfectly symmetrical. The idea of an inflation of the cosmos right after the Big Bang might provide a plausible explanation for this phenomenon.

"When a symmetrical vacuum changes into an unsymmetrical one—the process that happened right after the Big Bang, according to experts—a huge amount of energy is freed. This provided the melting heat of the vacuum. Cosmologists surmise that our matter, as we see it, originated in this transition. The particles that make up galaxies, stars, and planets are the residues of that transition from one vacuum state to another, shortly after the Big Bang, which broke the symmetry. It created the masses of elementary particles in the process.

"Should it be proven that this idea, commonly accepted by most cosmologists, is the correct one, then we ourselves—the stuff of which we are made—would turn out to be products of the vacuum. The vacuum, you see, has had an interesting career as our natural sciences developed. It began as an expression for the 'nothing,' for empty space without any properties. Today, the vacuum is a physical phenomenon that must be seen as the most important and most interesting concept in all of physics.

"Thank you for your attention."[2]

CHAPTER 6

What Is Mass?

I am one of those who spent a lot of time in deep
thought, but did not learn a thing in the process.
—Albert Einstein[1]

After his lecture in Berlin, Haller met with Einstein and Newton in
the living room of the house in Caputh. This time, Newton started
the conversation:

"There are a few things I would like to comment on concerning
your lecture yesterday, Haller. Also, I have a few questions—but let
them wait. Just tell me one thing: the success of today's physicists
notwithstanding, don't you think you are assigning too big a job to
the vacuum? For me, the vacuum used to be total emptiness; a nothing or, if you wish, a thing without properties. Well, maybe I was
mistaken. According to your lecture, the vacuum is the most complicated nothing imaginable, full of those strange virtual particles
and replete with a whole library of books on the laws of Nature. In
the end, you'll be claiming that gravitation is also a property of the
vacuum."

EINSTEIN: He not only claims that, this is exactly the truth. You'll
be astonished, dear Newton, once we get into the middle of this,
regarding gravitation.

HALLER: Take it easy, gentlemen, or we won't get anywhere. First,
Sir Isaac, back to what you said. I'll admit that, from your vantage
point, it is not easy to accept—that "empty" space is actually
equipped with a lot of physical qualities. But considering the understanding we have today, we have no other choice. Let's look, for
instance, at the creation of a muon and its antiparticle when an electron collides with its own antiparticle or, if you wish, through the
annihilation of this latter particle pair. This can happen provided

the energy is high enough. The mass of the muon is about two hundred times that of the electron. The muon, together with its antiparticle, will be created at the point of collision—a point where there was nothing before. How can it be that the muon is created with exactly the mass that we know of? It doesn't matter where it is created—whether on Earth or very faraway in the Andromeda galaxy. The muon will always be pulled out of the "nothing" with its given mass of 107 MeV; and there will be its antiparticle, with the same mass. Who on Earth is going to tell the muon, as it pops out of the vacuum, what mass it should have?

NEWTON: You might be right—it is pretty strange, this spontaneous creation of matter out of the nothing. One might think the vacuum "knows" beforehand what will be happening there all of a sudden.

HALLER: We can't get away from this: the laws of Nature are fully contained in the vacuum—just the laws, not matter itself. It is these laws that determine how matter will behave.

EINSTEIN: You think the vacuum somehow knows what exactly the mass of a muon—or of any other elementary particle which results from a collision—should be? That seems peculiar—a law is a law; that means it is something qualitative, you either have it or not. But the mass of the particle is just a number. Just those 107 MeV of the muon mass, for instance. Our world would hardly look different if the mass of a muon were, say, 87 MeV instead, would it?

HALLER: You bring up a difficult point. We don't know to this day what exactly is the mechanism that determines the mass of a particle. Yesterday, I spoke of a hypothetical field that pervades the vacuum everywhere, and that is responsible for the creation of all masses. As I said, it's just a hypothesis. We still don't know the details of the mechanism that is responsible for creating mass. But I imagine that this mechanism permits the muon mass to be just 107 MeV, and not 87 MeV or whatever. In other words: a world where the muon mass is 87 MeV does not exist—the laws of Nature do not allow it.

EINSTEIN: When you spoke about that strange mass field, I had to think back instinctively to what scientists used to call the **ether**, and of the unfortunate fate it received with my theory of relativity. Maybe your mass field will have a similar fate—but that is just a guess.

HALLER: Not at all! I would have no objection if you could propose a better mechanism for the generation of masses.

EINSTEIN: Hmm, you are not exactly modest with your demands.

NEWTON: Let's assume this mass field really exists. How can we then prove it experimentally?

HALLER: We assume that the mechanism that conjures up the mass of particles is itself characterized by an energy in the region of a few 100 GeV. That is comparable to the mass of the t-quark. In the simplest case, we imagine there is a real particle that represents the "mass field," that "carries" it.

EINSTEIN: You mean that these particles represent the "mass field" in a similar way that photons represent the electromagnetic field?

HALLER: Yes, but this time we are dealing with a particle that assigns to all other particles their given masses as it interacts with them. By the way, something similar happens with the photon: it interacts with all electrically charged particles, say, with the electron. Strictly speaking, the electrical charge of a particle is nothing but the permission that that particle may interact with the photon. And the strength of this interaction describes the magnitude of this charge.

EINSTEIN: I realize this, but I see a major difference between the electric charge and the mass of a particle. The electric charge always comes in steps—let's say it is zero, as for neutrons, and it is one unit of an electron's charge, or an integer multiple of that charge for other particles. It cannot be smaller. That's the way they serve beer in Bavaria: either you take a liter jug, or they won't serve you. But what goes for beer jugs does not go for mass. Electrons and muons, for example, have the same electrical charge, but their masses have this odd relative ratio of 1 to 207.

HALLER: I can't contradict that the mass of a particle seems to be something qualitatively different from its electric charge. Nevertheless, it could be that the "mass field" exists along with its carrier particle. This particle has its own name. It is usually called the Higgs particle, as I mentioned yesterday. This particle itself should have a given mass, too. Well understood, it not only has a mass, it also assigns a mass to all other particles.

NEWTON: A strange logic—first we want to explain mass by interaction with a field, then we say that the particle belonging to the field also has mass; I have to say, that's pretty odd, we explain mass by mass! That is like Baron Münchhausen of German folklore, who pulls himself out of a quagmire by his own bootstraps, not heeding my law of gravitation. If you ask me, that's a pretty dubious explanation.

HALLER: You can't dismiss it that easily, Sir Isaac. The mass field is a field, and also the mass of the Higgs particle is something very special. All other masses are just consequences of the mass of the Higgs particle, if you wish. The assumption is that this mass is somewhere in the range between 100 and 1,000 GeV. The particle would therefore be more than 100 times heavier than the proton, but hardly more than 1,000 protons.

EINSTEIN: So it would not be lighter than the Z particle, of which we have already spoken. Since we observe this particle in collisions of protons and antiprotons, we might assume that we can find the Higgs particle in the same way.

HALLER: In principle, yes. But the energy of our accelerators is, as it seems, not yet sufficiently high, neither at CERN nor at Fermilab in the United States. For that reason, a new proton accelerator is being built at CERN—the LHC (Large Hadron Collider). It is being installed in the tunnel of the LEP accelerator. With the help of this LHC, we'll be able to accelerate protons up to energies of about 7,000 GeV. Then we can collide them. Sometimes, if not often, a Higgs par-

FIGURE 6.1 A model of the LHC accelerator in the LEP tunnel at CERN. (*Photo:* CERN)

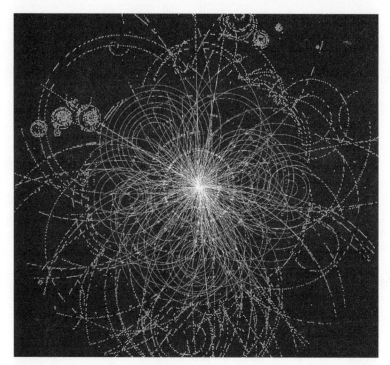

FIGURE 6.2 A computer simulation of a proton-antiproton collision at the LHC. Among many other particles, a Higgs boson is produced, decaying immediately into two Z particles. The Zs in turn decay into electron or muon pairs (a total of four particles indicated here by straight tracks), which are easily detectable. It is hoped that the Higgs particle—should it exist at all—will be detected in this way. (*Photo:* CERN)

ticle will be produced in the process, and we should then observe it indirectly from the particles into which it decays. You see, it is not a stable particle: it decays right away.

EINSTEIN: During these collisions, we deal with enormous energies, which means many particles can be produced. How then can we detect this "mass field?" That is surely not easier than finding a needle in a haystack.

HALLER: It's not quite that bad. The Higgs particle interacts with particles we already know, and it acts on these with a force that increases with their mass.

EINSTEIN: Ah—I get it. Since the mass of the Z particle is so enormous, the Higgs particle will decay into a Z preferentially; and the Z particle is easy to recognize.

HALLER: Indeed, we expect that the Higgs particle often decays into two Z particles and nothing else; but that can happen only if its mass is at least twice that of the Z, i.e., at least 180 GeV. Since every Z particle can decay into either an electron-positron pair or into a muon-antimuon pair, we have to look for collision events in which electrons or muons are emitted with large energies, and where the sum of the energies of the different particle pairs make up the energies of two Z particles. Technically, this is possible. And so at the LHC we hope to find the Higgs particle and any other particles that might have something to do with the mechanism of mass creation. In case the Higgs particle has a mass that is less than twice the Z particle's mass, there are other decay possibilities we can experimentally observe. But these are details we need not discuss here.

NEWTON: Well, it seems to me that the likelihood of this curious mechanism for the production of masses being correct is, in my opinion, pretty small. Anyway, it does explain the phenomenon of mass simply by the magnitude of the force with which the Higgs particle interacts with another particle the mass of which we want to understand. But that doesn't get us very far. We still don't understand why the muon is about 200 times heavier than the electron, or why the proton is about 2,000 times heavier. I sincerely hope that this curious mechanism for the production of mass is not, as Einstein would say, the "real thing"—the real reason for the phenomenon of mass.

HALLER: I don't want to contradict your skepticism, Sir Isaac. Maybe you are right and today's physicists are following the wrong track. By the way, there *is* a second way of producing mass that has no direct relation to the Higgs field.

EINSTEIN: Aha—there we are. I was sure there must be something else.

HALLER: We are talking about the mass of the proton—or, if you wish, the mass of atomic nuclei.

NEWTON: These, after all, are the masses that are most important for gravitation. Almost all the mass of a body is due to the atomic nuclei; less than one tenth of a percent of the mass has anything to do with the electrons. So where does the nuclear mass come from?

HALLER: When I said earlier that a deeper understanding of the phenomenon of mass is lacking, I was exaggerating. This is not the case for the proton or the neutron and, therefore, for atomic nuclei.

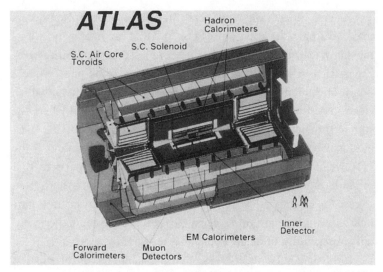

FIGURE 6.3 A schematic view of the ATLAS detector, one of the two large particle detectors being prepared for use at the LHC accelerator. (CERN)

The proton is not an elementary particle—it is made up of smaller objects, which we already introduced as quarks. We need exactly three quarks to build a proton. There are very strong forces that act between the quarks; they are carried by the particles of this force, which we call **gluons**. The proton is really a complicated dynamical system consisting of quarks and gluons. It looks rather like a small soccer ball, filled with quarks and gluons. It has a radius of about 10^{-13} cm—that is, about a hundred thousandth of an atomic radius.

NEWTON: That means quarks and gluons are the actual building blocks of the atomic nucleus; all the dynamics of nuclear physics can then be explained in terms of the physics of quarks and gluons. That is analogous to atomic physics, which can be similarly reduced to the dynamics of atomic nuclei and the electron shell that surrounds them. Isn't that so?

HALLER: Absolutely. But that is not our task right now—we don't have to get into nuclear physics at this point.

EINSTEIN: Still, let me ask you a question about this: if I correctly understand you, the quarks in the nuclei are, in some way, analogous to the electrons in the shell that surrounds them, and the gluons are analogous to the photons. The gluons are the "glue" that binds quarks into nuclei; the photons bind together electrons and nuclei to make up atoms. But electrons have a mass. Do quarks and gluons also have a mass?

HALLER: This is simple as far as the gluons are concerned. They are massless, just like the photons. The main difference between photons and gluons is that gluons can also interact with each other; the force between the gluons is as strong as the force between a quark and a gluon.

EINSTEIN: I get it—these intergluonic forces are obviously a major factor in the dynamics between quarks and gluons. These forces do not exist for the photons: light interacts primarily with matter, not with light.

HALLER: Thank God—if such a force existed, the world would look quite different. A laser beam could not exist if individual photons attracted each other. The Sun's rays would interact with each other and might not even get to Earth. In short, we have every reason to be grateful that there is no force that acts directly between photons. Now back to the quarks: I have already mentioned that they have a mass; just recall the t-quark. The quarks that make up the proton have a mass that is small when compared with the proton's mass, and can even be neglected in good approximation. Recall that there are other quarks in Nature that do not make up the proton but are the building blocks of heavier, very unstable particles.

EINSTEIN: What do you mean when you say you can neglect the quark masses that make up the proton? If I really did that, the proton would consist of massless quarks and gluons. Where then do we get the mass of the proton—which is, after all, almost 1,000 MeV?

HALLER: Now we get to the heart of the matter. Today's theory of quarks and gluons is as clearly formulated as the theory of **electrodynamics**; and that theory does provide the basis of all of atomic physics. It says that the proton is made up of massless quarks and gluons. The proton gets its mass from their interaction—the mass is a purely dynamical quality. You might call this a dynamical mass generation.

NEWTON: I like that. If I look at a proton more closely, I should be able to see massless quarks and gluons inside it racing around at the velocity of light. I can then imagine that the mass of the proton is simply due to the fact that prevailing forces lock the quarks and gluons into a small space. The mass of protons would then be noth-

FIGURE 6.4 Three quarks constitute the elementary building blocks of the proton. They are bound together by the exchange of gluons. The gluonic forces are the strongest forces we know in Nature. They keep the quarks from showing up individually, in unbound states, and have not been observed as single particles.

ing but the energy of the motion of quarks and gluons inside the proton.

HALLER: We can describe it this way, yes. But today, in particle physics, we don't speak in terms of the motion energy of the quarks and gluons; rather, we speak of their field energy.

EINSTEIN: Let me add a point here. If I understand you correctly, Haller, I have to imagine the proton as a bundle of quarks and gluons. In a way, they make up a ball of lightning that has a given field energy. This energy then makes up the proton mass according to my old equivalence $E = mc^2$, that now relates the energy of this ball of lightening to the mass of the proton.

It reminds me of my youth. When I derived my old formula $E = mc^2$, I took a field of electromagnetic radiation and demonstrated that it behaves like a material object which has a mass that is given by the total energy divided by the square of the speed of light. Here, things are obviously pretty much the same, and that makes sense. We finally have a situation where we can see explicitly how a mass comes about. I get it now: the "real thing" in the masses of atomic nuclei is the field energy of their quarks and gluons. With this we understand rather well the origin of the nuclear masses. I, for one, can live with that quite happily.

HALLER: Unfortunately, I have to revise this picture a little bit. Thus far we have not considered the masses of the quarks explicitly. If I now bring the quarks' masses into the picture, things change a bit. The masses of the quarks that make up the proton are on the order of 5 MeV—and with three quarks to the proton, that means only a few percent of the proton's mass has anything to do with the quarks' mass.

EINSTEIN: If my hair were not already gray, it would not turn so because of this few percent. Qualitatively, nothing changes. More than 90 percent of the mass of the atomic nucleus, and therefore almost all the matter that we observe in the universe—like stars, planets, interstellar gas clouds, and the like—can be understood dynamically. All this does not need your Higgs particle. It's just the missing few percent that is due to the Higgs field. Agreed—I will buy that. But I will admit that I do not have a good feeling about it. The mass of a rock of, say, one kilogram could be divided into 980 grams of dynamical mass from the quarks' and gluons' interaction, and only 20 grams from the Higgs mass. Doesn't that seem odd to you, Haller?

HALLER: You don't have to harp on that, Professor Einstein. I am not a friend of splitting up masses in this way; but I cannot exclude it either. After all, it is similar to the case of the atom. The main part of the atomic mass is that of the nucleus, and a small part comes from the electrons. That means we can split up the mass here. Today, we have concrete ideas how to describe the mass of electrons, quarks, and also W and Z bosons by examining purely dynamical effects. But nobody knows whether these effects are properly described. To find this out, we will need the new LHC accelerator being built at CERN, as I have previously mentioned. Note, please, that we are now talking about the forefront of today's physics research.

EINSTEIN: Fine, so we'll just have to be a little patient. I believe we should postpone further discussions about the mass problem until more is known. When we next meet—say, in the year 2010—we'll get back to this point . . . Now let's finally get to our principal topic, to gravitation. But everything at its proper time. I just remembered that I reserved a table at the hotel on Lake Schwielow; we'd better get over there quickly.

And that is what happened. The three physicists started walking toward the lake. Einstein knew a shortcut and took a narrow path that passed by a little pond. It did not take them long to reach the Haus am See, a hotel and restaurant on the banks of Lake Schwielow.

CHAPTER 7

Gravity—Is It a Force?

There are some experiences nobody can truly
understand unless he has lived through them: a
searching for truth that is filled with yearning and
with premonitions; the back and forth between
supreme confidence and utter dejection; and finally,
the breakthrough to the truth as it stands revealed.

—Albert Einstein[1]

After a copious meal at the restaurant Haus am See, during which
they enjoyed some excellent fish from the nearby lake and two bot-
tles of the local white wine that not even Newton had been able to
bypass, our three friends started happily on their way home. They
decided to take a little detour, along the lake's shore, that passed
through a peaceful village.

Among many important works, Theodor Fontane, a German
writer and novelist (1819–1898), also wrote descriptions of hikes he
took through the picturesque landscapes of what is called the Mark
Brandenburg. Einstein quoted from Fontane's *Walking Tour
Through the Mark Brandenburg*: "Lake Schwielow is a broad, comfortable, and sunny lake, possessing all the agreeable features of a
generous Nature." He added that there is an additional deeper reason for this: the Havel River, after leaving Potsdam, passes through
Lake Templin; it then has to squeeze through the Caputh juncture
and finally mellows down when it enters the comfortable basin of
Lake Schwielow. That is why Einstein liked this lake in particular;
whenever possible during his summer stays in Caputh, he would
take his sailboat down to Lake Schwielow.

Our threesome now wandered back through the woods, passing
two small ponds on the way to Caputh. They came to a clearing and
decided to take a little rest. Einstein walked over to a tree that offered
him a branch at the right height to do a few pull-ups, and finally
dropped down to the ground.

Newton watched in amusement, then picked up their previous discussion: "I guess you are trying to make sure that the gravitational force is still switched on, Einstein."

EINSTEIN: You might be surprised about what I am going to say right now. There really is no such thing as a gravitational force. That means there is also no such thing as switching that force off.

NEWTON: I take it for granted that you have read my *Principia*, am I right? If so, you should know the nature of gravity: it is a universal, all-penetrating force. And it proved that by pulling you back to the ground just now.

EINSTEIN: I don't dispute that gravity penetrates everything and is universal; but is it a force? I did believe that at one time, but then I had doubts. It was in 1907, while writing a review article about the special theory of relativity. In doing so, I realized that I could describe—in the framework of my theory—all the known laws of nature and all the forces, including electrodynamic ones. But I was unable to do the same with gravity. And do you know what brought me on the right track toward a solution of the gravity problem? A **gedanken experiment**—that is, a thought (or imagined) experiment—not so different from the experiment in which you just saw me fall down from that branch. I was sitting at my usual desk chair in the Patent Office in Bern when, all of a sudden, it struck me that the very weight of my body, which was pushing down the chair, would be unnoticeable to me if I were in free fall. I talked to an artisan who had fallen down from a tall scaffolding while performing some roof repair work in the center of town. His accident did not hurt him seriously. But he was able to tell me that he had no feeling of his own weight, of a gravitational force pulling on him while he fell down.

HALLER: The phenomenon of weightlessness that Einstein is talking about is well known to anybody who watches television these days. There, you can see the awkward maneuverings of astronauts in a space capsule, as they operate while experiencing weightlessness in the craft that is moving around Earth in our planet's gravitational field: but they do not feel gravity.

NEWTON: Very well, then—I did think there was something wrong with our idea of mass.

HALLER: What do you mean when you say that you thought something was wrong?

FIGURE 7.1 The astronaut and physicist Sally Ride is shown weightless in space. With her left arm she touches the "floor" of her space capsule, over which she would otherwise float freely. (*Photo:* NASA)

NEWTON: There are two different kinds of mass, as you know. If I pick up this rock and accelerate it as best I can, I feel a resistance. The rock does not want to be accelerated—it resists. This resistance is due to the mass of this rock—or maybe I should say, its inertial mass.

But the rock is also the source of a gravitational field. It acts on all bodies that surround it, especially on Mother Earth. That results in the attraction the rock exerts on the Earth or vice versa, the attraction of the rock by our Earth. Both of these are equal, and are the same thing. Now let's suppose I let go of the rock and it falls down to the ground: that is how it demonstrates the action of the gravitational force. The weight of the rock, therefore, is a measure of the strength of the gravitational force. And at this point, it has nothing to do with the inertial mass of the rock. Therefore, we should differentiate inertial mass from weight.

For the electrical force, this would be obvious. An electrically charged steel sphere that is in an electrical field of force is subject to the electrical power effect, but has inertial mass. The strength of the electrical force is given by the charge of the sphere; the strength of inertia by its inertial mass. Analogously, it would make sense to measure the strength of gravitation of the weight of a steel sphere by a kind of gravitational charge; and that is, you see, the "gravitational mass."

It is only astonishing that the gravitational mass and the inertial mass are precisely the same—or, more correctly, that they are truly proportional to each other. Suppose I take two steel balls and compare them, finding that the inertial mass of one of them is precisely twice that of the other. I will now measure the force that one of the spheres exacts on the other; and that, too, is twice as large as for equal spheres. Now you know the consequences: *all physical bodies fall toward a center of gravity with equal velocity, regardless of their masses.* The gravitational force pulling on the larger sphere is twice as strong, but so is that sphere's inertia twice as large; and that is why both of them fall down equally fast. **Galileo** remarked on that centuries ago.

If you read the *Principia* carefully, you'll notice that the equality of these two basically different kinds of mass made me wonder. Both are as different as apples and pears, and still, they are exactly proportional for most purposes. There is no way to register a difference between these two concepts. Well, I did not fall for that. I even conducted a few experiments. The equality of the inertial and gravitational masses sees to it that the period of motion of a pendulum does not depend on the material it is made of. A grandfather clock with an iron pendulum will indicate the same hour as another one that has a copper pendulum—just as long as the two pendula have equal lengths. I compared the periods of oscillation for wood and gold pendula. Later, I did the same thing with silver, lead, and glass pendula. Finally, I managed to put together pendula with masses that were mostly water.

I finally reached the conclusion that inertial and heavy masses are equal within a precision of at least one in a thousand. My original hope, to see a deviation, did not come to pass. I began to wonder why inertial and heavy masses are proportional to one another.

HALLER: You were not the only one to wonder about the equality. Heinrich Hertz, who discovered the wave nature of electromagnetism, thought about these things after you did, toward the end of the nineteenth century. The great Viennese physicist Ernst Mach and, subsequently, his Hungarian colleague Roland von Eötvös built special scales to investigate the equality, or equivalence, of inertial and heavy masses—and they reached a precision level of five in a billion.

Not before the 1960s was it possible to improve the accuracy of Eötvös's experiments. Robert Dicke of Princeton University put together a fancy setup that permitted him to prove the equality of acceleration of two different materials, aluminum and gold, in the gravitational field of the Earth, within his very small measurement errors of 10^{-11}. Both masses proved to be the same. This is one of the most precisely ascertained quantitative comparisons in the natural sciences.

NEWTON: With a precision of one in a hundred billion? Unbelievable—and even at that level, there is no noticeable difference? The two masses are truly identical, as if there were a secret principle that enforces this equality. Why then, in God's name, do all objects fall down in a gravitational field, irrespective of their masses? That can't be coincidental. What do you think, Einstein? Does that make sense to you?

EINSTEIN: My theory of gravity gives an answer for that. But to do this point justice, we have to go step by step. After all, it took me ten years until I found an answer that was satisfactory to me.

Also, as you know, my special theory of relativity is based, just like your mechanics, on one basic fact: *two systems of reference that*

are relative to one another, either at rest or in a state of uniform
straightline motion, are completely equivalent. Equivalent, that is, as
long as they are inertial systems. Such a system is clearly indicated
when an object moves in a straight line, and at constant velocity, in
the absence of external forces. Strictly speaking, such a system cannot be realized on Earth or in the vicinity of Earth because of our
planet's ubiquitous gravitational pull. It is most closely approximated in outer space, far from any celestial bodies—which might be
realized in a spacecraft somewhere, say, halfway between our Milky
Way and the neighboring Andromeda galaxy.

NEWTON: Even there it is only approximate. A spacecraft that
wanders across the universe, far from planets and stars, will still be
influenced a little by distant celestial bodies. But I admit, the farther
we move away, the better it is. It will never be possible to erase totally
the bothersome influences of gravitation. In other words: inertial
systems are basically nothing but figments of our imagination.

EINSTEIN: Surely, like many concepts in the sciences, the inertial
system presents an idealized limiting case. In reality, it can be realized only approximately. But still, a car that moves on a freeway at
constant velocity is a good approximation for an inertial system as
far as its motion on the road is concerned, disregarding the effects of
gravity.

NEWTON: Do you recall our detailed conversation regarding my
old concepts of absolute space? Although your special theory of relativity puts my concepts into the wastebasket of the history of
physics, I can't abandon altogether the idea that there is something
very special about space—or, if you wish, about spacetime. After all,
we can use all kinds of reference systems for the description of natural phenomena, not just the inertial systems. The latter, obviously,
stand out. Once I see one of those, any other system that moves uniformly and along a straight line with respect to the first one will also
be an inertial system. If you will allow me, I would like to introduce
at this time a new version of my absolute space, so to speak, an
absolute spacetime.

EINSTEIN: I think I know what you are going to say: you'll just
propose that absolute spacetime is nothing but the entirety of inertial systems.

NEWTON: Exactly, every single one of the infinity of inertial systems is like one particular language in which we can express physi-

cal phenomena. If you switch to another system, you only change the way you describe it; that's almost like saying we can make the same argument in English, in French, or in German. Reality is not affected by the language we use to talk about it. At every point of the universe that is far removed from celestial bodies, I can introduce such a system. This very fact is a hint that spacetime, just as your special theory of relativity implies, is not some arbitrary construction; rather, that it is characterized by its very simplicity.

HALLER: I would like to make a point here: the fact that all inertial systems are equivalent does not apply just to mechanical processes; it is also true for the rest of physics—for the electrical and magnetic phenomena. And it is true, as we know, for the speed of light, which is the same in any frame of reference—about 300,000 km per second, or 186,000 miles per second. This leads us right away to the special theory of relativity. Another consequence is the fact that there is no way to measure velocity in an absolute way. It is always relative, never ever absolute. An astronaut flying in a closed space capsule through the universe is not able to find out, from observations within that capsule, what velocity his capsule has. He has no way to find out. He would first have to tell us with what frame of reference he defines his velocity—relative to the Earth, to the Sun, or to our own galaxy?

This does not apply to accelerations. As soon as the astronaut switches his spacecraft's engine on, there will be new inertial forces that will press him against the wall of his craft. It will cause loose objects to fly around inside the capsule. Every automobile driver experiences the same effect when accelerating or stepping on the brakes. In an accelerated system, bodies do not move uniformly in straight lines. Now, their paths will, in general, be curved. The forces that are due to the acceleration are easily measurable. They have a meaning that does not depend on a reference system. We can use them to measure the acceleration they are to effect. The astronaut, when he estimates the force that pushes him down into his seat, is able to estimate the prevalent acceleration. But there is nothing he can say about the velocity. Acceleration is absolute, velocity is relative.

NEWTON: That is exactly what I meant when I talked about absolute spacetime. The latter must exist, because accelerations are absolute, while velocities are relative. When I wrote about absolute acceleration in my *Principia*, I came up with an example, Einstein, which also contains a hint that the concept of absolute space—or, more precisely, of absolute spacetime, must make sense. Here it is: Let's look at a water bucket, half filled. If I start rotating it, the water surface will become concave, its surface will rise toward the rim of the bucket. Now I imagine that I am an observer who rotates with the bucket. For this observer the bucket is at rest. He is probably astonished that the water level is not flat. But as soon as he finds out that his entire universe, together with the surface of the Earth, is rotating about him, he will conclude that his water bucket is not at rest after all, but that it is rotating. The whole process is happening in a rotating reference system; and this reference system is not an inertial system.

EINSTEIN: That sounds fine, Newton. I suggest we engage in two simple gedanken experiments. Let's take two space capsules without windows that we supply with all kinds of different physical devices. There is an astronaut in each one of them. The two have radio contact and can compare notes on experiments they perform. We place one of the capsules far away in outer space but leave the other one on the surface of our Earth. Now let's start the first experiment.

Both astronauts conduct experiments on free fall. The astronaut in outer space will observe that there is no gravitational force in his system; his reference capsule is an inertial system. All objects are either at rest or in uniform motion. A wrench he releases from his hand will stay wherever he places it, even in midair. The astronaut in the capsule that remained on Earth will observe that any object is being accelerated downward. His reference system experiences a gravitational force; it is not an inertial system. His wrench will fall down. Now let the astronaut in outer space start his engine and make his capsule move with an acceleration that corresponds precisely to the gravitational pull any object experiences on the surface of the Earth. Recall that acceleration implies the addition of a velocity of 9.81 meters per second, for every second it adds. What will the astronaut notice?

NEWTON: The astronaut won't be in an inertial system anymore. His wrench will "fall" down; more precisely, the wrench will move in the direction opposite to the acceleration at hand. But in reality, the wrench doesn't actually move; it is simply not being accelerated. At least not as long as it is not forced to do so by the astronaut, who might hold it in his hand as he starts the engine. In the absence of such a constraint, it will stay where it is, while the astronaut and his spacecraft will move away in accelerated fashion.

EINSTEIN: The wrench then "falls" down. Now, Newton, are you getting the meaning? The astronaut suddenly experiences a feeling of "weight," just like his colleague on Earth, or if you wish, like I did sitting in my desk chair at the Patent Office in Bern. If the astronaut did not know that he is in outer space and that he just started his engine—let's say he has been asleep ever since he left Earth and is waking up just as his foot accidentally sets off the engine of his spacecraft—he would have no way of knowing that he is no longer on Earth. In other words: we have created gravity artificially simply by accelerating our reference system.

NEWTON: And you really believe that the astronaut cannot find out if he is in an accelerated system or on Earth?

EINSTEIN: How should he? He would have an easy time if there were a difference, even a very small one, between the inertial mass and the weight of an object. Then the astronaut on Earth would, with the help of a scale, measure the gravitational mass; the other would measure the inertial mass. But since there is no difference between the two masses, there is no way to find out who is in the gravitational field and who is simply in an accelerated system. Or can you tell me another way?

HALLER: There is one thing Einstein has not mentioned specifically. So let me do it here. What we just now noticed is a kind of equivalence between an acceleration and a gravitational field. Einstein claims that all experiments proceed in precisely the same fashion, no matter whether they are performed in an accelerated spaceship out in the universe or in a capsule at rest on the surface of the Earth. He did not say that this applies not just to mechanical experiments but also to experiments with electromagnetic phenomena; it is true for what happens to light, to radioactive substances, even in chemical reactions or biological processes. He raised this to the level of a principle. This is what we now call the *equivalence principle*. It is the basis on which the general theory of relativity is built.

NEWTON: Wait a minute, does that mean both of you are now trying to tell me that we can simulate the presence of a gravitational field simply by performing an acceleration? Are you saying that acceleration and gravitation are totally equivalent?

EINSTEIN: Not only that, they are completely identical, at least locally—such as in a limited region of space, like the inside of our spaceship. I can easily see that this thought goes against your grain, my dear Newton. But basically, I have only taken the consequences of our notion that the two kinds of mass that we discussed are the same.

That reminds me of the ideas about the ether: you recall that experiments in the late nineteenth century showed there is no manifestation for an ether in which electromagnetic phenomena propagate. I put an end to that when I stated that not only is there no such manifestation, but there simply is no such substance as an ether at all. The special theory of relativity was the consequence of that argument, as you know.

A similar case can be made with the way gravity behaves. The equality of inertial mass and weight tells us that gravity and inertial mass appear to be identical. But I say this is not just an appearance— they *are* the same. Inertia and gravity are two sides of the same coin. And the consequence of that is the general theory of relativity.

In a gravitational field, all objects fall equally fast, Newton. And that is what points us in the right direction. For three hundred years, the indications were there, ever since you published your *Principia*. Nobody took your hint seriously. But I will say that is precisely what should have been done. Your hint indicates the direction in which we have to go if we want to solve the puzzle of gravity.

NEWTON: This, to me, is a hard nut to crack. And you will forgive me if it induces some digestive difficulties, as far as I am concerned. Gravity is not something you can switch on or off, as you might do with a light switch. But that is exactly what you are doing here. You claim that, in the interior of the accelerated spaceship, there is a gravitational field, as far as the astronaut is concerned. But for us, as observers from outside the spaceship, we see clearly that this is an illusion, because the astronaut is in accelerated motion together with the spaceship. That means that a gravitational field does not exist in the spaceship, and is at most an illusion of it. I would say that you are both guilty of confusing illusion with reality.

EINSTEIN: Before we continue to argue, let's engage in another gedanken experiment: We take our space capsule, place it on top of a tall tower, and let it fall down. In order for the capsule not to be smashed on the ground, we ignite its engine shortly before it reaches

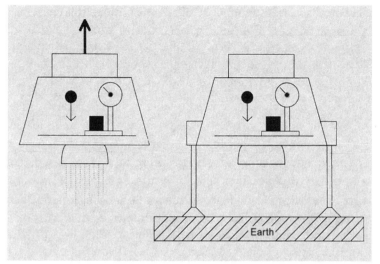

FIGURE 7.2 A gravitational field and an accelerated reference system are identical. At left, we show a spaceship in outer space that is being subjected to acceleration *g*. At right, a similar spaceship is shown to be at rest on Earth. The two systems are indistinguishable. Massive objects inside either system fall down by their weight.

the ground, so that it can touch down softly. Let's consider the process from the viewpoint of an observer on the ground who's watching the space capsule fall from the tower. Our observer will notice that the capsule is attracted by the Earth's gravitation, and that it moves in accelerated downward motion, just like the astronaut in the capsule and all the objects in it. Shortly before the capsule reaches the ground, the engine ignites; there is now a strong acceleration upwards, braking the fall of the capsule. Shortly thereafter, it comes to a stop.

NEWTON: Let me describe how this appears to the astronaut from inside the capsule: As long as the spaceship is on the tower, he registers a gravitational field around him, which is directed downward. At the instant when the capsule starts its motion downwards from the tower, the gravitational field will be compensated for by the acceleration. And at this point, there appears to be no gravity inside the capsule. The moment the engine is ignited, a strong gravitational field seems to appear, which is stronger than the usual gravity on Earth. After a short period, this field disappears and the Earth's usual gravity takes over. I stress the fact that I say that a gravitational field only "seems" to appear. In reality, there is none. This is an effect simulated by the acceleration.

EINSTEIN: No, Newton, I will not let you get away with that. This last statement is not correct. You said: in the free fall, there seems to be no gravity in the capsule. But I tell you: inside the capsule, gravitation does not only seem to vanish—there really is no gravity whatsoever. Through the effect of acceleration, gravity was eliminated, gone. Gravitation is a phenomenon that depends on the reference system. In one system it exists, in the other it does not. Seen in that way, gravity is not really a force, it just gives the appearance of a force. There is no way for the astronaut to know in a free-falling spaceship that, in fact, he is dropping downward in the gravitational field of the Earth. To him, that field does not exist. To wit: something you cannot measure does not exist.

NEWTON: Now stop it! You say that in a free-falling system there is no gravitational field. Does that mean that all experiments that the astronaut will perform inside his capsule will have the same result that his colleague in outer space will find inside his own inertial system? And I am now talking about truly all possible experiments—the electromagnetic or chemical experiments as well as the biological ones.

HALLER: Einstein in fact maintains that there is no way to tell the difference between the two systems. The free-falling system is also an inertial system. Basically, this is a fine thing: we stressed repeatedly how difficult it is to find a real inertial system, since the effects of our ubiquitous gravity cannot just be switched off. Einstein's principle of equivalence gives us a simple solution for this. Even here on Earth, we can, in principle, set up an inertial system—as our example of the free-falling space capsule proves. It doesn't even have to be a space capsule. An elevator plunging down in free fall is also such an inertial system—but, of course, only as long as it actually falls freely. Some difficult experiments can be performed only in a gravity-free space. We can help ourselves these days to occasions where airplanes actually do fall freely toward the Earth in a way that gives us true weightlessness for a few minutes.

EINSTEIN: We can solve this problem, in principle, by digging a hole through the center of the Earth and up to the opposite surface. A cabin that oscillates between both surfaces would then actually

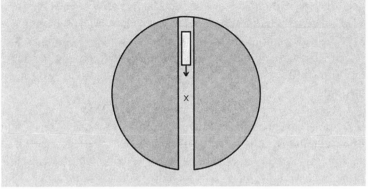

FIGURE 7.3 A cabin moves freely in the Earth's gravitational field inside a shaft that goes through the center of the Earth. The cabin oscillates back and forth between the shaft's two openings. There is no gravitational field apparent inside the cabin. The system is an inertial one, even though it is in accelerated motion with respect to the Earth. Effects of air friction have been neglected.

qualify as an inertial system, provided we could neglect the effects of friction and air resistance.

NEWTON: That is strange—here we have a system that is subject to constant acceleration, and still you call it an inertial system. If you define such a system on the grounds that any object will *internally* remain at rest unless exposed to some additional force, then you are right. Then it is indeed an inertial system. But you'll forgive me, gentlemen, I will need a while to reconfigure my view of our world, after the cracks it just suffered.

I propose that we stop our discussion for now. Let me just ask one more question: given that we permit accelerated reference systems to appear as inertial systems, due to Einstein's principle of the equivalence of gravitation and inertia, we are now close to admitting essentially all reference systems as such—no matter how extravagantly they move about in space. Now, your special theory of relativity is valid only for the old inertial systems. By that I mean real inertial systems, way out in free space, far removed from all objects that might exert some gravitational pull. But if we now admit all kinds of reference systems, what does that do to your theory?

EINSTEIN: That is an excellent question, Sir Isaac. Obviously, we need to find a way to describe the physical phenomena in all reference systems, not only in the old inertial systems. In particular, we need to generalize my theory of relativity for all reference systems. That is exactly what I set out to do in 1907. It is easily said but hard to do, as I soon found out. Only in the year 1915 did I find out what gravity really means. It is actually possible to generalize all reference systems: it can be done in my general theory of relativity—the principle of the equivalence of gravity and inertia, it turns out, is also a theory of gravitation. But it will take us a while to make that clear.

CHAPTER 8

Does Light Bend?

The surrender of fundamentally elaborated notions
on space, time, and motion may not be seen as vol-
untary, but rather as mandated by observed facts.
—Albert Einstein[1]

In the afternoon, the three physicists met on the large terrace. Einstein had managed to find an old blackboard, and placed it on a tri-
pod by the window. They had some coffee, and Haller started the
discussion: "Earlier today, we mentioned that, in the framework of
the special theory of relativity, we can define an absolute acceleration but not an absolute velocity. Acceleration is easily measured,
but an absolute velocity is certainly not. If we now also admit accelerated reference systems, or all imaginable reference systems, this
would mean that we have to give up the idea of absolute acceleration, too. The principle of the equivalence of gravity and inertia
forces that on us; after all, when we move within a gravitational
field, there is no way that we can measure absolute acceleration.
Recall the elevator that is falling freely. There is no acceleration
noticeable inside it. And when we do measure acceleration anywhere, we do not know if we are in an accelerated system or in a
gravitational field."

NEWTON: I still cannot believe that, from now on, the concept of
absolute acceleration shall be lost in the wastebasket of the history
of physics. Relative velocity, absolute acceleration—these are basic
properties of the special theory of relativity. As recently as this
morning, we discussed that we can take advantage of these features
so as to define a kind of absolute spacetime. And now you ask me to
give it up? Moreover, I don't see how a gravitational field that we
know can be eliminated by some appropriate acceleration—as the
equivalence principle apparently requires. You may not be able actu-

ally to touch a gravitational field; still, it is a real physical phenomenon, and we cannot just order it out of existence. What is worse, you tell me it can't be done by a purely kinematic operation, a subjective one to boot—simply an acceleration. A gravitational field, Professor Einstein, is not simply geometry, it is physics.

EINSTEIN: Maybe physics and geometry have something in common, Sir Isaac. By the way, no one said that by going to an accelerated reference system, the entire gravitational field—let's say the Earth's—can be eliminated. That can happen only as what we like to think of as a *local* phenomenon, like in our falling elevator, the dimensions of which are minuscule when compared with our Earth or its gravitational field. Let's say we increase the dimensions of the elevator in our gedanken experiment by a lot—we make it a linear tube of 100 km: in that case, the curvature of the surface of the Earth will become noticeable. We might then eliminate gravity in the middle of that tube, but not at its outer rims. To put it into fancy words: we can switch off the gravitational field in a given location only. It can be done "locally," but not "universally."

In that respect, there is a difference between a real gravitational field like the Sun's, and an artificially prepared field that is due to our passing from an inertial system to an accelerated system. That is what we did in our example of the accelerated space capsule, inside which conditions prevail as they do on Earth. Objects drop down to the floor, etc. The latter can be totally, and not only locally, eliminated when we change over to a system that is no longer accelerated.

NEWTON: You mean that a true gravitational field, like that of our Earth's, can influence the geometry of space, or if you wish, of spacetime? I like that—I could live with that. We could have gravity issuing from the structure of spacetime—wow! To do that, our gravitational theory must include the property that, at any given point, we can eliminate the gravitational field by a cleverly chosen reference system. That might be true.

HALLER: Newton, I am amazed at your ability to get to the right point. I think Einstein came to his conclusions pretty much in the same fashion.

EINSTEIN: It became clear to me around 1911; a correct theory could only be found if you start in reverse: you look for a theory of gravitation that does not follow directly from the equivalence principle. Rather, we build up a theory that, in the end, winds up implying that principle. True, it took me a very long time to find the appropriate train of thought. It would have taken even longer had I not gotten some vital assistance from a friend of mine who is a mathematician. I had to learn that the geometric concepts that were needed to realize my idea had already been developed by nineteenth-century mathematicians such as **Carl Friedrich Gauss** and **Bernhard Riemann**. In the process, I found that the geometrization of gravity was necessary, and that the equivalence principle leads to the phenomenon that light can be deflected by a gravitational field.

NEWTON: You mean that a beam of light from some distant star will, as it passes by the Sun, be influenced by the Sun's gravitational field—that it will be attracted as though it were a comet passing by the Sun? That doesn't sound unreasonable; after all, a light beam means energy, and energy means mass, as your theory of relativity tells us; and mass is subject to gravitational pull. I would like to point out that, in my book *Optiks*, I mentioned just such a possibility.

HALLER: There is one thing I would like to insist on: a light beam *always* takes the shortest path between two points. In normal space, that is a straight line. If we find out that a light beam is curved by a gravitational field, that implies that the shortest line between two points in space is not a straight line. And that means: something happened to the prevailing geometry.

NEWTON: You can't impress me with that, Haller. At first sight, this might seem odd. But a simple example from geometry shows us that things work differently in curved space as compared with normal space. And by the latter I mean space that we are accustomed to, and that does not have curvature. But when we look at the surface of the Earth, it is, to a good approximation, the surface of a sphere, and as such, a two-dimensional space. The shortest line connecting Berlin and London now turns out to be an arc, and not a straight line. If gravity does curve space in some fashion, we have to expect that our light rays will follow this curvature. It is just devilishly hard to work out the details of such a picture.

EINSTEIN: You can say that again. I must admit that I made a lot of mistakes trying to figure this out. Finally, in November 1915, I found the trick that will explain it. But we'll get to that later.

To illustrate the effect of gravitation on light, let me suggest another little gedanken experiment. This time we don't need a spaceship; a simple elevator in a skyscraper will do. First, let's consider the elevator at rest on, say, the tenth floor. It is subject to the Earth's gravitational field. Now we let an observer inside the elevator do an experiment. We make him drill a hole into one side of the elevator, so that light from the outside will enter. Then we have him do the same, at the same height, through the opposite side of the elevator. We'll then have him mount a strong light source outside the first hole so that it can send a beam of light—of laser light for best results—into the elevator. The beam shall be exactly horizontal. Now we will make the experimenter do something that is possible only in a gedanken experiment, not in reality: we'll have him switch off the gravitational field for a short while; this means that, in so doing, he creates the conditions that prevail far away in outer space. His reference system is now an inertial system.

NEWTON: I understand—now when he turns the light on again, the light comes through the first hole and leaves through the second, since the latter is exactly across from the other.

EINSTEIN: Now comes the $64,000 question: what will happen if he switches the gravitational field on again?

NEWTON: A question well put is half the answer. Had you asked me an hour ago, my answer would have been that the light coming in through one hole obviously will leave through the other one. But now, after our discussion about curvature, I'm not so sure anymore. I might surmise that the gravitational field will affect the path of the light in some fashion. Why else would you have mounted that gedanken experiment?

I suggest we first do another such gedanken experiment, but an easier one. I'll switch off gravity as before, but then I put the elevator into accelerated upward motion—supposing we can speak of "upward" in the absence of gravitation. As soon as the elevator moves upwards, I'll let the light beam enter through the first hole. It will cross the elevator and reach the opposite wall. If the elevator were at rest, the beam would exit through the second hole. But since the elevator moves upward in accelerated motion, the light will miss the second hole, hitting the opposite wall below that hole. The observer in the elevator sees that the light beam does not move in a straight line in his system, but along a curved path. He takes a closer look and notices that that path is a parabola, even though the curvature is exceedingly small. Now get this: this curvature of the light's path is apparent only to the moving observer. Another observer who remains at rest outside the elevator and looks at the light beam's path from that perspective will see the light propagate along straight lines. The curvature is strictly a consequence of the elevator's accelerated motion.

EINSTEIN: You can guess our next step: let's leave the elevator at rest and switch gravity back on.

NEWTON: Since this, according to your principle, is equivalent to an acceleration, the light has to move along a curved path. You win, Einstein! I agree with you—light beams will follow curved paths in a gravitational field. But I'm not overly astonished, since more than three hundred years ago I speculated on this. True enough, I did approach it from a different vantage point.

HALLER: We can do the gedanken experiment we just discussed in another way, without having to switch gravity on and off, which, as we know, cannot really be done. Instead, imagine I drop the elevator at the very moment the light beam enters it. The elevator will move downward in accelerated motion. We'll then bring it to a stop, so that our experimenter has a chance to escape unscathed.

NEWTON: Ah, I get it. Since gravity and acceleration are equivalent, the elevator constitutes an inertial system after we switch on

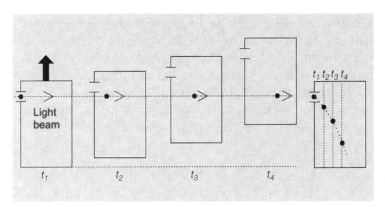

FIGURE 8.1 A light beam moves in a straight line in weightless space, as seen by an observer at rest across from a box that is in accelerated motion. We show the position of the light beam at given times, t_1, t_2, t_3, t_4. In the accelerated reference system, light beams move along a parabola with a curvature that increases with the acceleration of the box. (This effect is exaggerated in the illustration.)

the light. We have created the same conditions that would prevail in the absence of gravity. In other words: the light beam should move across the elevator to the hole on the other side and exit there. That was not possible when the elevator did not move. But now, the light beam arrives an instant later, and in this very instant the elevator has fallen down a bit.

EINSTEIN: Let's imagine the elevator is made of glass, and somebody looks in from the outside, observing the laser beam. What will she see? For her the elevator moves downward in accelerated motion. The second hole has moved down as the light beam hits the opposite wall. If the light exits through the second hole, it has to have "moved down" as far as the elevator. That means its path is curved by the gravitational field. The acceleration of the light beam in the direction of the center of the Earth is identical to the acceleration of a rock that we drop down: it amounts to the usual 9.81 meters per second squared.

HALLER: I owe you a cautionary remark. The effect we just mentioned is really minuscule. If we take the distance between the two holes to be three meters, the light will take one hundred billionth of a second, or 10^{-8} seconds, to cross the elevator. During this tiny moment, the elevator falls down a very small distance, of course— to be precise, less than 10^{-14} meters, or about as much as the diameter of a large atomic nucleus.

EINSTEIN: I was aware that this distance is measurable only in principle, but not in reality. Still, when we observe a distance on Earth of 3,000 km, light will need about 0.01 seconds to cross it. In this time, the light beam will be pulled "down" about half a millimeter—certainly a tiny amount; but still, this amount can be measured. If we do the experiment on a faraway planet, on which gravity is 1,000 times stronger, this distance would be one half of one meter—and this is very easily measurable.

In 1911, I followed the argument we just discussed to calculate the deviation from straightline motion which is experienced by light that propagates to Earth from some distant star and, in the process, passes close by the Sun. It amounts to about 0.84 seconds of arc. Again, this is a small amount, but certainly a measurable one during a total solar eclipse. I wrote a letter about this to the astronomer George Ellery Hale, director of the Mount Wilson Observatory in the San Gabriel Mountains above Pasadena. I never received an answer.

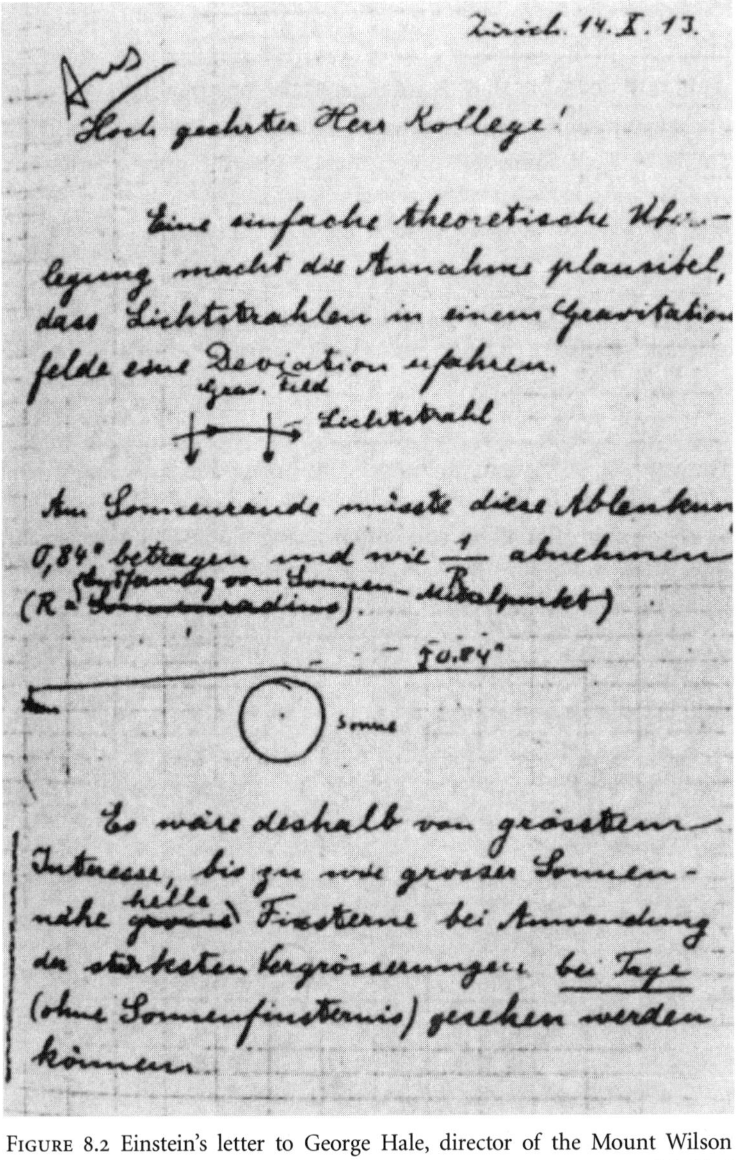

Zürich. 14. X. 13.

Ans

Hoch geehrter Herr Kollege!

Eine einfache theoretische Überlegung macht die Annahme plausibel, dass Lichtstrahlen in einem Gravitationsfelde eine Deviation erfahren.

Grav. feld

Lichtstrahl

Am Sonnenrande müsste diese Ablenkung 0,84" betragen und wie $\frac{1}{R}$ abnehmen (R = ~~Sonnenradius~~ Abstand vom Sonnen-Mittelpunkt).

0.84"

Sonne

Es wäre deshalb von grösstem Interesse, bis zu wie grosser Sonnennähe ~~grosse~~ helle Fixsterne bei Anwendung der stärksten Vergrösserungen bei Tage (ohne Sonnenfinsternis) gesehen werden können.

FIGURE 8.2 Einstein's letter to George Hale, director of the Mount Wilson Observatory, concerning the bending of light beams in the gravitational field of the Sun. (Albert Einstein Archives, Jerusalem)

HALLER: By the way, here is a quote from your paper that you published in the *Annals of Physics* in 1911. It ends like this: "It would be highly desirable that astronomers take up this question, even though my consideration presented here might look insufficiently proven or even somewhat adventurous. After all, quite apart from any theory, we have to ask ourselves whether we can actually establish the influence of gravitational fields on the propagation of light with today's experimental means." Seen from today's vantage point, these words were truly prescient: they were written at the outset of a decade-long experimental investigation into the effects of general relativity, albeit their prediction was a bit tentative at the time.

EINSTEIN: Thank God I was a little imprecise at the time—four years later, I knew better. I had committed a conceptual error. The actual value of the deviation was bigger by a factor of 2: it amounted to 1.7 seconds of arc. It is quite interesting to look at the reason for this discrepancy, and we'll get back to it. But it has nothing to do with our fundamental considerations above. They were correct as we stated them. Still, subsequent developments that led up to the experimental determination of this deviation became quite dramatic.

HALLER: You will forgive me if I interrupt you here. I studied this matter in detail a little while ago. So let me give you a short summary.

EINSTEIN: Fire away, Haller. I must admit I might not recall the details.

HALLER: The first preparations for the measurement of the bending of light began about 1913. A solar eclipse was to occur in August 1914—to be precise, on August 21, 1914. But nothing could happen experimentally, since that was the start of World War I. Then, in 1917, English astronomers pointed out that the solar eclipse expected for May 29, 1919, would be an excellent date to observe this bending of light, since our Sun would be on the path between the Earth and the Hyades cluster in the Taurus constellation, which is in the middle of a segment of our skies that has lots of very bright stars. But the region of the total eclipse was a bit below the equator; that meant expensive expeditions would be needed to do the measurements correctly. Professor **Arthur S. Eddington** from Cambridge University was the director of one of the two research groups organized by the Royal Greenwich Observatory. He was a strong supporter of the general theory of relativity. He chose the volcanic island Principe in the Golf of Guinea, West Africa. The other expedition went to Sobral in the northwest section of Brazil.

After months of meticulous work on their observations, the actual value for the light's deflection was finally determined in the fall of 1919 and made public on November 6 of that year at a meeting of the London Royal Society and the Royal Astronomical Society. Einstein's prediction was confirmed within the range of experimental error.

The president of the Royal Society, the physicist Joseph John Thomson, gave a talk that ended like this: "What was discovered here is not some isolated island; rather, we are dealing here with a whole continent of scientific thought. This is the most important result concerning gravitation that we have had since Newton's days; it is therefore only proper that it should be announced in a meeting of this Society with which he was so closely connected. The result is one of the greatest achievements of human thought."[2]

Newton walked toward the window after this quotation and gazed out over the Brandenburg landscape. Everybody was silent. Finally, Newton turned around and said: "Congratulations, Einstein! You really hit the jackpot here. Now back to work. Experiment has had its say and has eliminated all doubts. Now let's look more closely at the theory. But not today—I need some time to think. Forgive me, therefore, if I retire to my room."

CHAPTER 9

A Flat World Curved

Logical simplicity is the only path that leads to more
profound knowledge.

—Albert Einstein[1]

It was dusk by the time the three physicists met again in Einstein's
house. The housekeeper had prepared a light meal, and they took it
on the terrace. After dinner, Einstein steered the conversation back
to the real subject of the discussion.

"The main problem that rattled me when I was setting up the
general theory of relativity was of a more formal—or, if you wish, of
a more mathematical—nature. When it dawned on me that gravity
is not a real force but rather an apparent force, somewhat like the
centrifugal force that is generated by rotational motion, I had to find
a way for a quantitative description that would show that this apparent force can be eliminated; we just have to switch to an appropriate
reference system. Now I had to pay dearly for not having taken my
math courses all too seriously, back in my student days at Zurich
Tech. My math professor **Hermann Minkowski** called me a lazy
bum; and right he was, given that I had mostly physics buzzing in
my head. Anyway, I wish I had done more math at the time. That
would have saved me many a headache, and probably a few years of
my life during which I worked out the details of the general theory
of relativity. So I suggest we first concern ourselves with a few details
that—seemingly—have nothing to do with the theory, but rather
with curved spaces. Sir Isaac, you'll be up for a few surprises: things
are not as simple as you saw them when you postulated that there is
such a thing as absolute space."

HALLER: So let's first look at a few special examples. The space we
are able to observe with our eyes has three dimensions. Each point

in space is defined by three numbers; we call them its coordinates—length, width, and height. Nevertheless, we can also imagine space that has fewer than three dimensions. The simplest of these is a space of dimension zero; that is a point, or a space, that has no extent at all. We can fix this point in our three-dimensional space anywhere, it is then a subspace of dimension zero inside our three-dimensional space.

NEWTON: I know which way you are heading—if I move this point through space along a straight line, which may stretch out to infinity, it will still have only one dimension: each point on this line can be described in terms of one number only, one coordinate.

HALLER: You deliberately chose a straight line; but I could also consider a circle or an even more complicated trajectory for our point. A circle is also a space, but this one has a special property—that the line it describes winds up finding its way back to where it started. It is a space in itself. It is not infinitely long; rather, it has a finite length given by its circumference. In contrast to the straight line, a circular line is a finite one-dimensional space—although it has neither start nor finish.

EINSTEIN: We can play this game even further. Let's take our straight line and rotate it around one point such that it creates a plane. Now we have a two-dimensional space. Each point in this space is described by two coordinates.

HALLER: I would like to stress that this plane, while it is a part of our three-dimensional space in which it is embedded, is an inde-

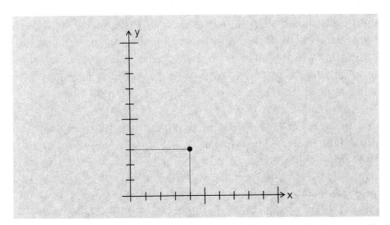

FIGURE 9.1 A plane is a two-dimensional space. Every point in this space is defined by two coordinates.

pendent space nonetheless. In other words: we might forgo this embedding, so that we are left with just a two-dimensional space, nothing else. The fact that we introduced it as part of three-dimensional space is unimportant for its structure.

EINSTEIN: We might imagine small, two-dimensional organisms—say, amoebas—and let them move in our plane. These creatures have a given length and width, but no height. They know right from left, front from back; but they have no idea of up and down. There is no third dimension for them, just as in ancient Egyptian paintings. Early Egyptians had no notion of perspective—they depicted everything, even people, in great detail, but in only two dimensions.

Looking at this flatworld from our three-dimensional viewpoint, we might amuse ourselves as those flatlings crawl about on their plane in blithe ignorance of any such thing as a third dimension. Never mind that this third dimension is above and below them.

HALLER: Surely there must be mathematicians in this flatland who wonder if one could not add a third dimension, at right angles to the other two. The plane would be embedded in a three-dimensional space. But you might not be very successful in convincing their fellow creatures that are accustomed to just two dimensions. The mathematicians might be told: what is this nonsense you are talking about? A third dimension that is at right angles to the other two? That is science fiction, it's a long way from reality. That third dimension would be just an idea, accepted only by a small mathematically trained elite of flatlings. The first ever to talk about the usefulness of a further dimension just might wind up burned at the stake, just like **Giordano Bruno** on the Campo dei Fiori in Rome.

The flatland-mathematicians might invent various peculiar geometric structures—unintelligible to the normal flatland-citizen. For example a sphere, the three-dimensional analogue to the circle; or a cube, the three-dimensional analogue to the square. In the plane, you might observe only a circle that tells you about the sphere, if the latter pierces the plane. If the sphere moves vertically with respect to the plane, it will touch this plane at one point; as it penetrates the plane, there will be an ever-increasing circle until it reaches the equator's size, only to decrease again. And finally, the sphere will vanish back into the third dimension.

EINSTEIN: We can continue this game further, but let's be careful: now we get into the same bind that the flatlings hit earlier. We get our three-dimensional space by enriching the plane with an added dimension. For us, this is easy to picture. After all, we live in three-dimensional space. But now what if our three-dimensional mathematicians come and say: this three-dimensional space will henceforth be supplemented by an added fourth dimension, vertical to the other three. Henceforth, our three-dimensional space becomes part of a bigger, four-dimensional one. This, of course, is nothing but mathematical fiction. Our three-dimensional people won't give a damn. But let's imagine the following scenario: strange four-dimensional creatures are observing us from the outside, from a fourth dimension. They might play all kinds of tricks on us. For instance, they might push a four-dimensional sphere through our space—say, the four-dimensional analogue to a soccer ball. We, in our limited three-dimensional view of things, would first notice just a point, then see it grow into a sphere, increasing in size to some maximum, only to vanish again. This could happen anywhere, even inside a bank's vault. For thieves, four dimensions would be ideal, first during the burglary, and later for hiding the loot.

HALLER: Easy now—our space has only three dimensions, thank goodness. We might ask why this is so. I have to admit that modern science does not provide an answer. But one thing has to be mentioned: we might ask how the laws of nature appear in spaces of higher dimension, that is, in a four-dimensional space. Recall, we know that the law that describes gravitation, just like that which describes electrical forces, depends on the number of dimensions. Let's look at the attraction between two electrically charged small spheres. The force between the two decreases rapidly as we move the spheres apart. It decreases with the square of the distance between the two. Suppose we increase the distance by a factor of two, the force reduces by $2^2 = 4$. If we now remove one dimension, we study the law of electrical forces in two dimensions and find a linear decrease. Note that this can easily be done in three-dimensional space, by measuring the attraction between two parallel rods that are electrically charged—and while we are all but ignoring the third dimension, we measure a linear decrease. The force weakens by a factor of 2 if we double the distance. Now we add a further dimension, in addition to the three already present, and again we look at the forces between two electrical charges as we vary the distance between them. We find that the force decreases by a factor 2^3, that, is by a factor of 8.

NEWTON: The attractive force decreases rapidly: the way you tell it, I can imagine there would be some problems. Anyway, it certainly would be a totally different world, and I would certainly have to rewrite my *Principia*.

HALLER: Indeed, there are serious problems. For instance, we could find that a rapid decrease of electrical forces as a function of distance would lead to instabilities in the orbits of charged objects in an electric field of force—say, of electrons as they orbit the atomic nucleus. The result might be that there are no longer any stable atoms. Not even planetary orbits would be stable in a four-dimensional world. So, it is a good thing that the space we live in has three dimensions and no more, but also no less. It is the best possible world as far as dimensions are concerned.

EINSTEIN: Luckily, we have three dimensions; I would get claustrophobic with only two. I couldn't even go sailing on Lake Schwielow with only two dimensions.

But now let's get back to our original topic—back to the plane. That is the simplest example of space as described by Euclid in antiquity: to be sure, it's a pretty boring type of space, without the slightest bit of structure. One important quality of this Euclidean space is the simple fact that the sum of the angles in any triangle must amount to 180 degrees. This is the seal of quality for Euclidean space.

A while ago, we constructed the plane by rotating a straight line about one point. Let me now replace the straight line by a circle and, again, I will make it rotate about its origin. In three-dimensional space, this makes us create a spherical surface. This surface, again, is a two-dimensional space—just like the surface of our Earth is two-dimensional. As the circle was before, we now can see the spherical surface as a finite, two-dimensional space; it is devoid of start and finish, it is closed up in itself. It has a further feature that the plane does not share: it has curvature. We are privileged because we can see from our three-dimensional viewpoint a spherical surface; and of course, we cannot help noticing the curvature of this surface. Straight lines from three-dimensional space cannot be embedded in this surface.

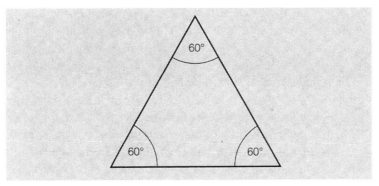

FIGURE 9.2 On a plane, a two-dimensional Euclidean space, the sum of the angles in a triangle always adds up to 180 degrees. We show an equilateral triangle with angles of 60 degrees each.

NEWTON: Anybody can tell that the spherical surface is curved; but this curvature is a concept that has something to do with this embedding we discussed. Unless we have this embedding in normal three-dimensional space, it makes no sense to talk of curvature.

HALLER: You bet! I'm glad you mentioned this point. Even so, we can speak of curvature itself. The German mathematician Gauss was one of the first who realized in the nineteenth century that we can study curvature as a feature of a plane without leaving this plane; without, that is, adding a further dimension.

NEWTON: We talked about those flatlings before. Now let's suppose we have such flatlings on the surface of our sphere. They might start to get enterprising, and engage in expeditions in all directions—just to find out what kind of world they live in. They would realize that an expedition they start in any given direction will, if this direction is maintained, bring them back to their point of departure. They would also discover that their space is unlimited but finite. This is an experience shared by explorers that sailed around our world some five hundred years ago.

Next, the mathematicians at a flatling university would announce that the surface of a hypothetical sphere in a three-dimensional space is one way to interpret their flat universe. The distance covered by an expedition, once it gets back to its point of departure, would be seen as the circumference of a circle on this sphere, the radius of which is the distance they traveled divided by 2π. All this is easy to understand; but how should these poor flatlings

FIGURE 9.3 An image of the German mathematician Carl Friedrich Gauss, who established important foundations of non-Euclidean geometry, can be seen on every German DM10 banknote (German Federal Bank).

learn anything about curvature in their space—that is, space they cannot leave—if they have no recourse to a third dimension?

HALLER: You just mentioned that the flatlings could find out whether their world is finite or whether it is infinitely large. That is also true about curvature. Gauss was the first to address this problem, and he came up with the right solution. Let's look at a triangle on our spherical surface: of course, this cannot be a normal triangle, the way we draw it on a sheet of paper—on a flat sheet, that is. Rather, we mark three points on our sphere and connect them by the shortest lines we can draw; each of these lines is part of a circle with the radius of this sphere. We call these lines that connect two points in the shortest possible way **geodesic lines**. It is an expression well known in the artful science of geodesy. We know that any airplane that flies from New York to Paris will normally follow one of these geodetic lines above the surface of the Earth. On a flat plane or in our normal three-dimensional space, the geodesic line is a straight line. For that reason, the geodesic line on the surface of a sphere is not just the shortest connection between two points; it is also the closest analogue to a straight line in any plane. It is the line we follow when we drive from one point to the other by just going straight, not turning the wheel to the right or to the left.

We can now easily see that the sum of the angles in a triangle that consists of geodetics is no longer 180 degrees—it must be a little larger. We can, for example, construct here on Earth an equilateral

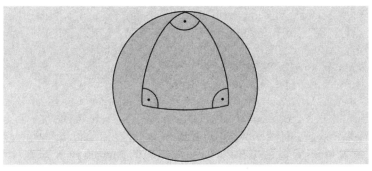

*A Flat World
Curved*

FIGURE 9.4 An equilateral triangle on the surface of a sphere, where the sum of its angles adds up to 270 degrees: two of its corner points are on the equator of the sphere, the third at its northern pole.

triangle that has one of its angles at the North Pole, with the other two at the Equator. We see right away that each of the three angles at the corner points turns out to measure 90 degrees. That means the sum of the three angles is 270 degrees; it exceeds the usual sum we know from a triangle in the plane by 50 percent.

NEWTON: Now I understand what you were driving at. The flatlings on a sphere could survey triangles which, for simplicity, we chose to be equilateral; now, just by adding up the three angles, they can find out whether their space has curvature. That really makes sense. Starting from your example, we can easily see that the sum of the angles on the surface of a sphere is not a constant; rather, it depends on the length of the sides of the triangle.

EINSTEIN: You are right. Haller gave us an extreme example when he spanned a triangle from the North Pole to the Equator. When we choose small triangles—and by small, I mean triangles that are a great deal smaller than the circumference of the sphere—the sum of the angles will be only slightly larger than 180 degrees. If we reduce the size of the triangle more and more, the sum of the angles will get closer and closer to the customary 180 degrees. The reason is that, for small triangles, the curvature of the two-dimensional space we are studying can be neglected. Suppose I am sitting on a given point of the sphere—say, at its North Pole: as long as I study the two-dimensional space in my immediate vicinity, I can forget about curvature altogether; for all intents and purposes, the space around me is flat. It is well described by the plane that touches my sphere at the North Pole—just like what the mathematicians call a *tangent*.

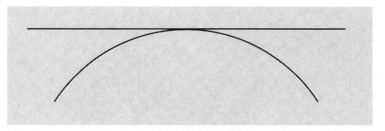

FIGURE 9.5 A tangent is a straight line that touches a curved line—in the case shown here, it touches a circle. In the same way, a tangential space (a plane) can touch a curved surface at any point. In the *immediate* vicinity of the point of contact, the curvature of line or surface can be neglected.

We can easily see that this tangential space can be applied to a curved surface at any point, and not just to the surface of the sphere. This tangential space, which is just an auxiliary concept, has all the properties of a plane, i.e., of Euclidean space. The sum of the angles for any triangle in this space will be 180 degrees.

NEWTON: I think I now understand what you and Haller mean. If I were a flatling on a spherical surface, totally ignorant of the existence of a third dimension, I would proceed as follows: first I would measure very small equilateral triangles and find that the space in my vicinity is Euclidean. The sum of the angles in our triangles always amount to 180 degrees within measurable error. Now I examine larger triangles and realize that the sum of angles is no longer 180 degrees but, say, 185 degrees. For even larger triangles, the sum gets larger, up to 270 degrees, as you said before. From these results, I can draw my inferences on the curvature of space; I can even find out that I live on a spherical surface. You convinced me, it can be done. It can be established whether a given space has curvature—even without looking at this space from the outside, without the help of further dimensions.

EINSTEIN: That is no coincidence. The curvature of a space is a property of that space; it has nothing to do with its being embedded in a space with additional dimensions. That was Gauss's important discovery when he worked out all all the details for curved planes.

For a flatling living on any curved plane, his world is not only a plane consisting of a collection of points; he also knows all relevant lengths and their relations. At any time, he can give the shortest connection between two arbitrary points on his plane. Those would be the lengths of his geodesics. He can easily measure these lengths.

That is why we speak of **metric** spaces—spaces where we can always tell the distance between two points. Our three-dimensional space is also a metric space. Everybody knows how we can get the distance between two points: it is given by the length of the shortest connection between the points, and that is a straight line.

NEWTON: I understand. This kind of a curved surface is a two-dimensional metric space—the distance between two points can always be measured. Suppose I draw a right-angled triangle and call its corner points *A*, *B*, and *C*, where the right angle is at corner *C*. In our Euclidean space, Pythagoras's rule applies: *for any right-angled triangle, the square of the hypotenuse is equal to the sum of the squares of the other two sides*; and thus, the sum of the squared distances between points $(AC)^2 + (BC)^2$ equals the distance between points $(AB)^2$ (i.e., the hypotenuse). But this is valid only in Euclidean space. Pythagoras's theorem applies only as long as there is no curvature.

EINSTEIN: Flatlings who, for starters, have not observed the curvature of their world, would learn in school about Pythagoras's rule as one of the pillars of their geometry. As soon as they find out about curvature, by doing precise measurements of the sum of the angles in triangles, they would have to accept that Pythagoras's rule is no longer strictly correct.

There are other properties of Euclidean space that we take for granted and that also do not apply in curved spaces; there is, for instance, the "fact" that parallel lines never intersect. Let's look at two points, *A* and *B*, on the Equator 100 km apart from each other,

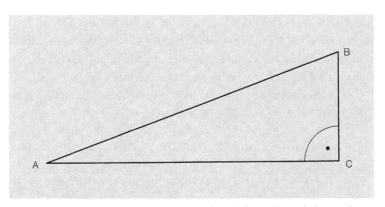

FIGURE 9.6 In Euclidean space, Pythagoras's rule for right-angled triangles—i.e., that $(AC)^2 + (BC)^2 = (AB)^2$—is valid.

in the middle of the ocean. From each point, a ship travels directly north along its own meridian. Both ships sail on parallel lines, but these lines will intersect, as we well know, at the North Pole. In a Euclidean space without curvature, this would not be possible.

HALLER: There is another well-known fact that is valid in Euclidean space but is no longer applicable in regard to curved surfaces: the circumference of a circle with radius R amounts to $2\pi R$ in a plane, but not on a curved surface, when we define the radius as the distance between a point on the circle and the origin in its center. It would be senseless to define anything else as the radius. Let's look at the surface of a sphere such as the Earth; then, let's just draw successive circles around the North Pole. If the radius of the circle is small in comparison to the circumference of the Equator or in comparison with the radius of the sphere, then the circumference of the circle, divided by 2π, equals the radius to a good approximation. But when the spheres get larger, this is no longer true: curvature will see to it that the distance between points on the sphere and the axis of the Earth, which must be the circumference of the sphere divided by 2π, is smaller than the distance between points on the circle and the North Pole. Recall that we measured this distance on the surface of the Earth.

The most extreme case is reached when we look at the largest circle possible on the sphere—the Equator. Here, the circumference is exactly 4 times as large as the radius. The circumference divided by 2π is therefore much smaller than the radius. This gives a good example for the importance of where we make our observations. An observer who sees the circle as a feature of three-dimensional space will, quite obviously, take the distance of points on the circle from the Earth's axis to be the radius of the circle. For the flatling who lives strictly on the surface of the Earth, this does not apply. He is forced to choose the larger radius R; the straight path to the axis of the Earth does not exist in his two-dimensional world.

NEWTON: I note then: as soon as there is curvature, the circumference of a circle divided by 2π is always smaller than the radius.

HALLER: This is not what I said. What you just told me is valid for the surface of a sphere or, more generally, for surfaces similar to those of a sphere. And that is what we call surfaces with *positive curvature*. There are also other kinds of surfaces where that is not the case. Take, for instance, the surface of a saddle. Imagine you are on a mountain tour and you just arrived on top of a saddle. In one

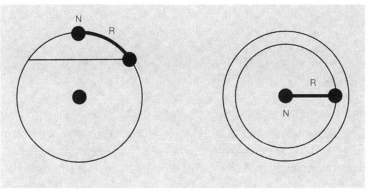

FIGURE 9.7 On the spherical Earth, we look at a sphere drawn around the North Pole at distance *R* as measured on the Earth's surface. The radius of this circle—by definition the distance of the points on the circle from the Earth's axis—is smaller than *R*.

direction—say, in front of you and behind you—the slope goes downward; at right angles to that direction, the path will rise. If you now draw a circle around the place where you're standing, just marking a closed line of points that are equally distant from you as you turn around, you will notice that the circumference of the line that you draw, divided by 2π, is larger than the radius.

NEWTON: I understand. In the case of a saddle, the curvature we observe in one direction is opposite to that we see in the other direction. Therefore, the circumference divided by 2π is larger here than in the plane. That is really interesting! We can divide all surfaces into those that have a curvature like a spherical surface, and others that look like a saddle. A planar surface is just halfway between these two possibilities.

EINSTEIN: That is exactly what we do. Surfaces that resemble those of a sphere have the same curvature in all directions; we call them surfaces with positive curvature. Surfaces that are similar to the saddle, we call surfaces with *negative curvature*. But one cannot necessarily always tell whether a surface has positive (i.e., spherical) curvature, or whether that property is negative, as in the case of a saddle. The curvature can change—it depends on the point we choose for observation: a surface that has positive curvature at one point may have negative curvature at another. Our example will show this very easily.

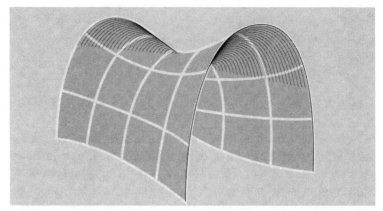

FIGURE 9.8 A saddle surface is a two-dimensional space of negative curvature. A circle around the middle of the saddle surface has a circumference that is bigger than 2π times the radius.

Let's look at two mountains that are connected by a saddle pass, as is so often the case. Let's assume we climb to the top of the mountain, passing over the saddle. It is clear that the curvature we see on the peak is positive, while the one on the saddle is negative. This means that, as I climb along a ridge from the saddle point to one of the summits, I walk along a line of points that pass from an initially negative curvature through a point of zero curvature and then on the peak, where the curvature is positive.

NEWTON: I take it that the curvature is a local phenomenon that depends on the particular point we look at. Only in rare cases, such as a spherical surface, is the curvature the same everywhere. In general, it varies from one point to another; that makes sense when we look at surfaces of various shapes.

HALLER: A mountaineer who looks at his path can, in principle, detect the amount of curvature and whether it is positive or negative; there are, of course, singular situations we have to accept, in dicontinuities like rock ridges or ravines. This means that the curvature is a function of the point in question. Curvature defines the structure of a surface and makes it much less boring, if you will, than a plane, where the curvature is the same all over. In the plane, after all, the curvature is simply zero.

NEWTON: Up to this point, we have discussed curvature in a graphic way. But now that we accept a change of curvature from one place to the next, I wonder how the flatlings that live on curved surfaces should be able to describe this. After all, they have nothing like the privilege we are accustomed to, being able to observe curvature in all its features. Sure, they can draw circles around themselves, wherever they stand, and notice curvature at any given point by studying the ratio of radius and circumference; that will tell them the curvature at any given point; but that is a pretty tedious procedure, for sure.

EINSTEIN: You are right. That's why mathematicians have developed a procedure that permits them to define curvature as long as they have a mathematical representation of the surface. We don't have to discuss the details just now; otherwise, we would get bogged down in mathematical details. But let me say this much: it is possible to describe an arbitrary curved surface by assigning a quantity to each point of the surface that fixes the distances on the surface. For instance, it will say the distance between this point and another one is so and so much. We call this set of quantities a metric **tensor**. It determines the metric structure of a surface. Let me pick a simple example to illustrate this: take our well-known Euclidean coordinate system. Every point in the plane is fixed by its coordinates x and y. If we know the coordinates of two points—say, (x,y) and (X,Y)—we know not only the positions of these two points, but also the distance ℓ between them. The square of this distance is given by the sum of the squares of the differences in the two coordinates:

$$\ell^2 = (x-X)^2 + (y-Y)^2$$

In this case, the *metric tensor* is particularly simple. Here it means nothing more than the mathematical prescription that tells us how to calculate the distance. That prescription is: *take the sum of the squares of the two differences of coordinates, and that gives you the square of the distance between the points.*

But nobody forces us to choose this definition. We might just as easily take another one:

$$\ell^2 = (x-X)^2 + 2(x-X)(y-Y) + (y-Y)^2$$

or, quite generally,

$$\ell^2 = A(x-X)^2 + 2B(x-X)(y-Y) + C(y-Y)^2$$

where A, B, and C are arbitrary constants. It is now only a small step to the statement that the quantities A, B, and C are dependent on the coordinates x and y—that they may have different value as we change x and y. This, however, means that the formula above is valid only for small distances, i.e., for the case where the distance between two points we are looking at is sufficiently small.

The metric tensor, mostly denoted by the letter g, is given by the quantities A, B, and C. Usually it is written in a square arrangement

$$g = \begin{pmatrix} A & B \\ B & C \end{pmatrix}$$

For the plane, the metric tensor is trivial—the quantities A and C have a value of one, and B is zero. This means that Pythagoras's theorem applies and that the distance between two points can be easily determined from their coordinates. It was to Gauss's merit that he noticed the fact that the distances on a curved plane can be described by a metric tensor; in this tensor, the quantities A, B, and C change from one location to the next. It is that simple: the transition from the plane to a curved surface is simply due to the fact that A, B, and C are now variables. The metric tensor denotes the length of a path on the surface; in the process, it defines the curvature uniquely.

NEWTON: But the connection between the metric tensor and the curvature is fairly complicated. First, I have to build up the surface and then, from its structure, I have to determine the curvature. Isn't there a faster way?

EINSTEIN: There is indeed: the curvature of the surface can be obtained through purely mathematical means from the metric tensor; that is done by the calculation of another mathematical quantity, which we call the *tensor of curvature*. Its exact formula is not of interest here. The important feature is simply that it can be calculated from the components of the metric tensor. The relevant mathematical formalism was developed in the first half of the nineteenth century by mathematicians such as Gauss, **János Bolyai**, **Nikolai Lobachevski**, and, in particular, by Riemann.

HALLER: We can imagine the tensor of curvature in the following fashion: when we hike into the mountains, we take along a map of the region. The surface of the Earth is clearly quite strongly curved in the mountainous region. Still, the map's design has to make do with a two-dimensional representation. The map as such is nothing but a two-dimensional Euclidean space. Now, if you want to find out the distance between two points in the mountains—say, the distance from some summit to my starting point in the valley—it is not sufficient if I measure the distance on the map and multiply it by the appropriate scale factor. This works well enough for a bicycle tour in the plains, but not in the mountains. As every mountaineer knows, we take resource to what is called lines of constant elevation; they connect, on the map, points that have the same elevation. Only when we include these lines of constant elevation in our calculations do we have a chance to take a good stab at the actual length of our route. That means we need these lines to find out the actually prevailing distances at hand. The lines of constant elevation then somehow stand for the tensor of curvature.

EINSTEIN: I would like to point out that the curvature of a given surface has a well-defined meaning for each and every point on it. It does not depend on the set of coordinates our flatlings might choose to describe the surface on which they live. When we have a curved surface, it makes no sense to try and use normal right-angled coordinates that we know from planar surfaces. True, a mountainous landscape is shown on a map that has only two coordinates, just like a plane; but this representation is an artificial construct which we get, for instance, by taking an aerial photograph of the relevant area. What we really get that way is the projection of the landscape onto a flat surface, with no information on relative elevations.

Any yachtsman who goes out to the open ocean will tell you that any point on the surface of the Earth—that is, on a two-dimensional curved surface—can be denoted by two sets of data. It takes nothing but the numbers for geographic longitude and latitude. Since the Earth revolves about a fixed axis, the North and South poles are singular points on the surface of the Earth. The largest circle we can fix on the Earth's surface around either the North Pole or the South Pole is the Equator, just one well-chosen circle of latitude. Parallels of latitude are then fixed, in contrast to meridians of longitude. The meridian denoted as the circle of zero longitude—that is, the international zero of longitude—was arbitrarily fixed so that it connects both poles with the location of the Greenwich Observatory near London.

The example of the Earth demonstrates that the description of points with the aid of geographic length and width is unique, but it is also arbitrary. Any other coordinate system might also do; we could, for example, choose as coordinates lines of arbitrary curvature on the surface of the Earth. The coordinate system we choose has no basic importance.

Another example is the saddle surface we mentioned before. It can also be described by a set of curved coordinates. The important point is that once we know the coordinates and the metric tensor, we can calculate all curvature properties on the surface, for every point. It is clearly and uniquely given and does not depend on the coordinate system.

NEWTON: Let's assume we describe such a saddle surface by two completely different coordinate systems. If I use these to calculate all properties of the curvature in this surface—that is, if I calculate the tensor of curvature in two different ways—the resulting curvature must be the same.

EINSTEIN: Precisely. *The curvature is independent of the coordinate system.* It is, if you wish, just a fact that can be described in two different languages. If I talk about the battle of Waterloo in English or in French, that makes no difference for its outcome: Napoleon is defeated either way. Coordinate systems are like languages, but the curvature is reality; the tensor of curvature can be seen as the text we use to describe the geometric situation.

NEWTON: I do think we have talked enough about two-dimensional curved space. It is now time to get to our main point: let's leave the world of the flatlings and move on to fully three-dimensional space.

But it is late now, gentlemen. Let me just wish you a good night.

CHAPTER 10

Curved Space and Cosmic Laziness

Sometimes I ask myself why it is I who put together the theory of relativity. It must be that most grown-ups never think about the problems of space and time—they are sure they did that when they were children. My own mind, on the other hand, developed slowly; only as an adult did I start to wonder about space and time. So it was probably a natural thing that I delved more deeply into these matters than normal children do.

—Albert Einstein[1]

Next morning, the three physicists took a cue from the lovely weather and left on the sailboat outing they had planned. Einstein had reserved a yacht down in the village the day before; so, right after breakfast they walked over to the pier by Lake Templin. A little after nine they set sail, and Einstein steered the boat westward in the direction of the Caputh Narrows, which joined Lake Templin and Lake Schwielow. Shortly after the yacht had gained some speed, they navigated in a southwesterly direction, toward the village of Ferch, and their discussion resumed.

NEWTON: I surmise that what we discussed yesterday about flatlings was meant to be nothing but an exercise, so that we could get ready to start looking at real three-dimensional space. Let me quickly recapitulate: we had seen that two-dimensional flatlings, with some knowledge of mathematics, are able to describe the curvature of their flat but curved universe in a clear fashion. They also realize that the curvature of their space is usually dependent on the location being measured, which they do so by surveying circles and their radii, or triangles and their individual angles.

It is remarkable that the properties of curvature become noticeable only when larger surfaces are surveyed. Over small distances, a curved surface always appears planar. Now I ask myself: What is it that makes up the curvature of my two-dimensional space, my surface? Could it be that its curvature is actually a physical property, a phenomenon that might be tied to other physical phenomena, such as the dynamics of matter itself? That would really be a major point—a synthesis of geometry and physics.

EINSTEIN: I am amazed to hear this from you, Sir Isaac. After all, it was you who proclaimed the idea of absolutely structureless space, of space with divine qualities, not to be interfered with, endlessly removed from physics: space that is as perfect as a flawless crystal, where straight lines go on perfectly in all three directions, in the most boring fashion. Geometry and physics as a unity? When I read your *Principia*, I found nothing of the kind.

NEWTON: You are right, but let me tell you that I pondered the problem of interaction between geometry and matter a long time before you. In my mechanics, this interaction is, of course, one-sided. Space acts on matter via inertia; but matter does not act on space. And that is a situation I always considered unsatisfactory and odd. The same goes for the equality of heavy mass and inert mass. In your theory of gravitation, this appears not to be the case. I am curious, Einstein, how you managed to get a grip on this one.

HALLER: Let me suggest that we advance systematically—and that, for starters, we don't worry about the reaction of matter in space. Let's first apply to a three-dimensional space what we did for our flat, two-dimensional space. The decisive idea of the nineteenth-century mathematicians—specifically Gauss and Riemann—was this: that the three-dimensional space, in which we live, can also be warped. That is to say, it can have curvature. This means that we can describe the properties of our space by a metric tensor that varies from point to point and thereby fixes the curvature.

NEWTON: Would you go so far as to say that we can directly observe curvature in our three-dimensional space, just like our hypothetical flatlings, by measuring triangles or circles?

EINSTEIN: Exactly. Gauss even gave this a try. He used the best geodetic instruments of his time and tried to measure a triangle from the summits of three mountains in the Harz Range. Above all, he wanted to find out whether the sum of the three angles actually added up to 180 degrees, as we would expect in Euclidean, structureless space. And he did not find a deviation from this. The sum of the angles was, within the large error of the experiment, consistent with 180 degrees.

Today, we know that the space close to the Earth is indeed curved because of gravity. But the effect is so small that Gauss had no chance to observe it. Still, he certainly was on the right track.

HALLER: Just a few mathematical details: the geometric properties of three-dimensional curved space—or, more precisely, the structure of its curvature—are described here just as for a two-dimensional surface, by the metric tensor. This tensor denotes, as we said, the metric properties that are given by the various distances in this space. Curvature is fixed by the tensor of curvature, which in turn can be determined from the metric tensor.

EINSTEIN: My dear friend—you and I are well acquainted with the technical terms of mathematics for non-Euclidean geometry, but Newton cannot possibly know them. Therefore, we should stay away from them whenever we can.

I would like to mention a specific case—the three-dimensional analogue of a sphere. A spherical surface, when interpreted as a two-dimensional space, is a curved surface with the specific properties that its curvature is positive and the same everywhere; it is given by the radius of the sphere. We can now ask: what form do I get when I take a constant positive curve in a three-dimensional space?"

NEWTON: I would expect to get something like the surface of a sphere in four dimensions. That means I have to imagine that I have to add an additional dimension; and in this hypothetical space, which I will call *hyperspace*, I will have the surface of a sphere. This new spherical surface is nothing but the totality of points in space that have the same distance from the origin—just like in a normal sphere in three-dimensional space. Since our sphere itself has four dimensions, the surface has three dimensions, i.e., as many dimensions as our space. If our space had such a structure, then the astronaut who flies in his spacecraft always in the same direction would finally reach his starting point—a kind of cosmic boomerang. The structure of this space would inevitably lead the spacecraft back to where it started its journey.

HALLER: Better still—if the astronaut were to keep a journal of his route, he could determine the radius of that shape, i.e., the radius of the curvature of his space. Let me stress again that the introduction of a fourth dimension—that is, the invention of hyperspace by way of the discussion of a four-dimensional sphere—is nothing but a formal mathematical trick. We don't need this fourth dimension; neither do we need it when discussing the curvature, because the latter belongs to three-dimensional space alone. This would be a self-contained space with positive and constant curvature. It will have a finite volume, but no fixed boundaries.

NEWTON: Let's assume our space is of that kind. Then we would have to be able to observe deviations from Euclidean geometry, at large distances, say, when we survey triangles. Gauss did not detect anything when he tried that. But what is the situation today? With the help of satellites, we should be able to get very precise surveying data of celestial bodies.

HALLER: This has been done, but we have to discuss the details later when we look at gravity. Let me just say this much right now: the space of our universe, as it seems, does not have the structure of the surface of a four-dimensional sphere. A large-distance curvature of space has not been found to date. If it exists at all, it must be very small.

NEWTON: But we looked not only at the surface of a sphere; we also looked at saddle surfaces that possesses negative curvature. Does that change anything? Is there a three-dimensional analogue for the saddle surface?

EINSTEIN: No problem. That would be a space with a constant, negative curvature; but in contrast to the spherical case, this space would extend out to infinity. An astronaut in this space would never get back to his point of departure.

It is interesting to observe a normal three-dimensional sphere in a spherical space and in a saddle space. The surface of a sphere in Euclidean space is 4π times the square of the radius. Just as the area of a circle changes when measured on a spherical surface, the relation between surface and radius no longer applies in non-Euclidean geometry. In principle, we could also determine the curvature of the space by measuring the surface of a sphere in a space and compare the result of our measurement with the quantity $4\pi R^2$. A deviation between the measured Euclidean radius and the radius as measured on the surface would be the extent of the curvature. If the surface is smaller than in the Euclidean case, we are dealing with positive curvature; if it is larger, we are dealing with negative curvature.

When we talked about light as deflected by the Sun, we were actually talking about the curvature of space that is expected in the framework of my theory of gravitation. On Earth, this amounts to very little—the deviation between what we would expect to be able to measure in an ideal case differs by only 1.5 millimeter from what we calculate from a survey of the surface. In other words, it is so tiny that we can all but forget about it. In the case of the Sun, there is a noticeable deviation, amounting to some 500 meters.

Hey, careful there, Newton! We are getting too close to the shore. I'd better turn around quickly.

Einstein gave some slack to the rigging; the sails flapped in the wind, and he managed a classic turn. The boat, which had barely missed running ashore near the village of Löcknitz, now took off in a southeasterly direction.

Newton: The three-dimensional space in which we live is also metric. If I disregard the effects of the theory of relativity such as the contraction of lengths, I have no trouble measuring the distance between two points. Wouldn't the metric tensor tell me about that?

Einstein: Of course it would—it knows all about the structure and the curvature of space. Just as in a curved surface, the shortest connection between two points is precisely defined—it is the geodesic line connecting the two points. At every point of an arbitrarily curved three-dimensional space, I can add a three-dimensional Euclidean tangential space, which, of course, will not have any curvature. Provided that you are only interested in the immediate surroundings, this tangential space is a useful approximation. At that point itself, real space and tangential space are the same.

Newton: In other words, the curvature of a three-dimensional space is, as with a curved surface, something we can determine only over larger distances. A very small spatial volume does not show any curvature. Rather, it looks like a small regular part of three-dimensional Euclidean space.

Einstein: Right you are. You get a curved space by simply joining many small Euclidean spaces in a slightly lopsided way. We are able to make a curved surface by joining together these many small flat surfaces.

You just mentioned the relativistic effects of the special theory of relativity. Indeed, we should not forget about them when we now

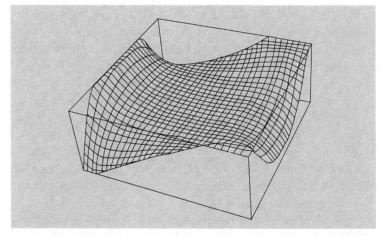

FIGURE 10.1 Example of a curved surface: in the immediate vicinity of each point on the surface, the curvature can be neglected. The surface touches a tangential plane that can be defined for every point. Therefore, the curved plane can be approximated by joining together many such tangential planes in neighboring points. (By permission of H. Mitter, Graz)

discuss gravity. We mentioned several times that the theory of gravity implies that normal three-dimensional space must appear warped. This, however, is just a side effect of my theory—it is unimportant for most purposes. We have to look at space and time

NEWTON: In the special theory of relativity, time is a further dimension that gets added to our three familiar spatial dimensions. Since you obviously put great emphasis on time in connection with gravity, I suggest we forget about two-dimensional space for the present. Let's just deal with only one spatial dimension and, of course, one time dimension. That gives us the advantage that we can make do with just two dimensions, so we can draw whatever we discuss, on paper.

HALLER: All right then, I will draw time as a vertical coordinate, at right angles to the spatial axis. We now have a two-dimensional spacetime structure. All the points that lie in a straight line at right angles to the time axis and intersect it at time *t* tell us about space at time *t*. A body at rest in this space will be represented as a straight line parallel to the time axis. It is the **world-line** in two-dimensional spacetime. In spacetime, we see that a body at rest does not appear as a point. In this two-dimensional space, it is not at rest; rather, you might say, it sails through time just as it ages. Being at rest in space

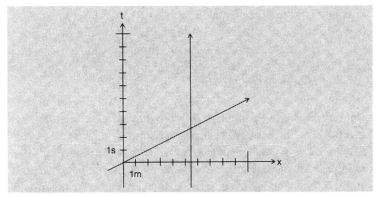

FIGURE 10.2 Two-dimensional spacetime. An object at rest is defined by a world-line that runs parallel to the time axis. In our example, an object is at rest at $x = 5.5$ meters (vertical line). If the object moves at constant velocity, its world-line is inclined with respect to the time axis. Our example shows an object that is at $x = 0$ when the time is $t = 0$, and moves away from this point at a velocity of 2 m/s (inclined world-line). The degree of inclination of a world-line describes the velocity. The greater the velocity, the stronger the inclination of the world-line is in the direction of the x axis.

certainly does not imply being at rest in spacetime—whatever rests will rust, as the Germans like to say. If the body moves at constant velocity, its world-line has an inclination relative to the time axis. If it moves at nonconstant speed, its world-line will show curvature.

NEWTON: All right, let's look at the world-line of an apple that falls from the tree at time 0 and hits the ground exactly one second later. My law of gravitational acceleration defines that the apple must start falling from a height of 4.9 meters to hit the ground one second later.

EINSTEIN:: Very well—I'll call the height of the apple at any given time $x(t)$. In the spacetime drawing, its world-line is a recumbent parabola. At time $t = 0$, the apple detached from the tree at a height of $h = 4.9$ meters. The height x decreases quickly, since the apple, accelerated by gravity, moves downward. After one second, the apple hits the ground; its world-line has arrived at the point $t = 1$ s, $x = 0$.

NEWTON: You just said that the apple was accelerated downward by gravity. But in your opinion, gravity is not a force but a consequence of geometry. Can you please explain to me how this apple's curved world-line, which we are discussing, can be achieved by geometric effects?

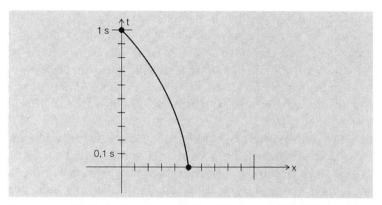

FIGURE 10.3 The world-line of an apple that falls from a height of $h = 4.9$ m to the ground, which it reaches after exactly 1 second. The world-line is part of a parabola open to the left.

EINSTEIN: That is precisely the same question I asked myself, right after defining the principle of equivalence in 1907. In order to answer to your satisfaction, we have to mention a few more aspects of spacetime. To start with, what about geometry in spacetime? I am not talking about the specific metric properties of spacetime. Our plane in spacetime consists, like any other plane, of points that are characterized by coordinates t and x. While a normal plane in space is nothing more than an orderly collection of space points, a plane in spacetime is made up of points in both space and time. Point x will be seen at time t. This we call an event—something happens at time t in point x. Our two-dimensional spacetime plane is a collection of events just as a normal two-dimensional plane is a collection of points.

NEWTON: If I remember correctly, one essential aspect of your special theory of relativity is the fact that it imposes a metric structure on spacetime.

HALLER: Seen relativistically, it makes no sense to say that this process takes a given amount of time; remember, that amount of time would depend on the frame of reference, because of time dilation. The length of a body also depends on the system. At high speeds, it will be foreshortened. In the special theory of relativity, this is a consequence of the fact that the velocity of light is the same in every system. But earlier we saw that there is something independent of the system—something absolute—the relativistic distance between two events. For a normal surface, which is described by two coordinates x and y, the square of the distance of a given point from the origin is arrived at by the sum of the squares of the two coordinates: $a^2 = x^2 + y^2$. In the event space of the special theory of relativity or, if you wish, in our event plane, the square of the distance between the origin of spacetime and any arbitrary event point is also given by a square of the coordinates. But this, as you remember from our earlier discussion, looks a little different.

NEWTON: I know, it is the difference of the squares of time and space or, more precisely: $a^2 = (ct)^2 - x^2$. Time does not appear as time in distance, but as time multiplied by the speed of light, which corresponds to the spatial distance.

EINSTEIN: This is necessary since, in the special theory of relativity, space and time are intermingled or, to express it in terms of geometry, are twisted into each other. That can be done only when space and time are measured in the same units, say, in meters. On the time axis, the quantity that corresponds to one meter is the time that light needs to propagate by one meter; that is the minuscule time of about 3.3 times 10^{-9} seconds. We could also retain units of seconds on the time scale, but that would mean we would have to use very large units on the space axis: a light second, which is the distance light travels in one second—is, to be precise, 299,792,458 meters.

Do you recall that we had to introduce this strange measure of distance to guarantee the constancy of the speed of light in every reference system? When I now change to a new reference system—in our case a spacetime plane—it means that we change to a moving system, by observing this event space from a moving train, you might say. We now get a new coordinate for time and a new one for space after this rotation. The transition to a moving system is therefore something of a transformation in spacetime and for that reason is sometimes called a *Lorentz transformation*, after my friend Hendrik Lorentz in Leiden.

The value of this concept of relativistic distance is due to the fact that this transformation does not change a thing. I can now say quite generally: looking at two events A and B in our event plane, I have no trouble in determining the square of the distance between them. It is $c^2(t_A - t_B)^2 - (X_A - X_B)^2$. This difference is the same in all reference systems. It is the quantity most important in physics.

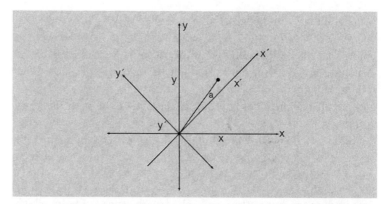

FIGURE 10.4 In a normal plane, the distance between an arbitrary point and the origin is given by the sum of the squares of the coordinates of that point. After a rotation (transformation) of the coordinate system (in our example by 45 degrees), this is true also for the new coordinates. The distance has not changed, it is "invariant under rotation."

Space and time are not important—what matters is the distance between the events.

NEWTON: Our two-dimensional spacetime has a metric structure, since the distance between two arbitrary points in space and time—that is, between events—is a given. However, I have to admit that this is a strange distance that I cannot imagine geometrically. It may be positive, negative, or zero—and this is clearly different from what happens to distances in normal space.

EINSTEIN: I'll agree with you there. Unfortunately, I cannot think of a suggestive graphic interpretation. When we imagine some given plane, we take it for granted that the distance between two points is nothing but just that, the spatial distance between the two of them—a quantity that is always positive, never zero or negative. Nature, however, does not care about how we picture this. In the special theory of relativity, this distance is just the difference between two numbers; as such, it can have all kinds of values. This, of course, is due to the fact that we are looking at space and time jointly in one coordinate system. In the special theory of relativity, time still has its own special role. It is not just the fourth dimension next to the three spatial dimensions.

NEWTON: I understand. Geometry in event space, where the distance between two events is described as you say, makes good sense.

It is a metric space; it is just that distances—or what we call the metric—are given in an unusual way. That is why it is advantageous to take recourse to the metric tensor. We can try all kinds of distances—we just have to find the physically relevant ones.

HALLER: The unusual definition of distances, by the way, leads us to another curious consequence of the theory of relativity. Let's look at the world-line of an object that is located at $x = 0$ and does not move. Its world-line is identical with the time axis. We now choose two events that are connected by this world-line: first, the origin, and second, an event with the coordinates $(t, x) = (1\,\text{s}, 0)$. The distance between these two events is given by just the time difference—that is, by one second to be multiplied by the velocity of light; so, this distance is about 300,000 kilometers. I can, of course, take all kinds of other paths in my two-dimensional spacetime and connect the two events by these. You can easily see that the "length" of any other path is smaller than 300,000 km, where I use the term "length" according to our definition of the distance between two points. For example, I could connect both events by a light beam. A light ray leaves one event and arrives after half a second at a point 150,000 km distant; it will be reflected there, so that it will reach its point of departure after another half second, having spent one second altogether to reach its starting point again. The "length" of this path is zero.

In general, this "length" of a path between two events, divided by the speed of light, is defined as the *proper time* of the path chosen. This definition is due to the fact that it would be shown by a watch that we send on a journey along the world-line considered. The time difference between the beginning and the end of the journey is what we call the proper time. It can take any time between zero or some maximum value, depending on which path it took—as a result, time is a quantity that depends on the path chosen between one event and another.

We can see that the straight world-line of our object at rest is identical with the line of maximum "length"—that is, the longest proper time. If I want to make sure that a clock that is at the origin at time $t = 0$ will be back at the same point at $t = 1$ second, and that will show the elapsed time as one second, the prescription is: leave the clock alone, don't move it! The slightest movement will see to it that the proper time elapsed will be less than one second, because of

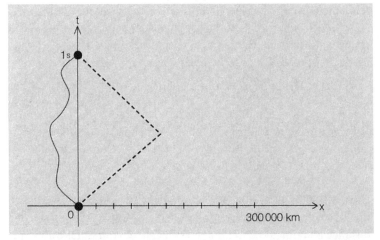

FIGURE 10.5 Different world-lines connect two events, the origin and the event (1 s, 0). The greatest length according to Einstein's definition, which is 300,000 km for this one second, is obtained when we join the two events by a straight line along the time axis. The dashed line corresponds to the path of a light signal which is emitted from the origin and is reflected after 150,000 km. The length of the world-line is zero. The curved solid line represents the world-line of an irregular motion. The length of this world-line is much smaller than 300,000 km.

time dilation. We see that in our spacetime plane things really look different. In a normal plane, the straight line between two points provides the shortest connection between them, what we call a geodesic line. In the spacetime plane, we are instead dealing with the longest connection. In the first case, we are dealing with a minimum, in the second case with a maximum of the connecting distance. Still, even in a spacetime plane we speak of a geodesic line. The straight world-lines of objects moving freely in two-dimensional spacetime are, as we see, the geodesic lines in spacetime.

EINSTEIN: You say that as though it were a matter of course, Haller. But let me stress again: *an object that moves freely in the universe and is not subject to an external force will follow a world-line from one event to the next such that this line provides the longest connection between these two events.*

My old friend **Bertrand Russell** thought up a nice expression for this. He called it the principle of "cosmic laziness." I can only agree with him. Nature, when interpreted relatively strictly, does not like to hurry. Here is another example for the relevance of treating time as one of the coordinates of spacetime.

When looking at normal space and an object that moves freely along a straight line, it follows a trajectory that provides the shortest distance between two points. This is not the case in the special theory of relativity. Instead of efficiency, laziness rules here. Nature always realizes the longest connection between two events.

But look! We've made it to the village of Ferch. I know a nice little restaurant on Main Street here. It's about noon—so let me suggest that we also follow the principle of cosmic laziness and close our discussion. After lunch, I have a special treat for you, Newton. You'll be amazed when we turn our attention toward gravity.

CHAPTER 11

Time Bent

There is something odd about scientific endeavor:
often enough, it's most important to see where there
is no need for a concerted effort. On the other hand,
it would be wrong to pursue goals just because they
are easy. We need to develop an instinct about just
how far we can get if we really try as hard as we can.
—Albert Einstein[1]

After a short walk through the village, our threesome returned to the pier. Haller assisted Einstein in setting sail. There was a lull, and the boat's motion was very slow. Finally, Einstein fastened it to a pole at the edge of a shoal. Not a word was exchanged on board. Haller recalled what Rudolf Keyser, Einstein's son-in-law, had told him about his father-in-law's sailing habits:

Sailing is what he considers his best relaxation. There is a landing just a few minutes from the house, and he keeps his mahogany boat anchored there. Einstein has no use for long outings by boat, or for attempts to set speed records. He likes daydreaming on board. He enjoys distant panoramic views, the effects of light and color on faraway shores. He delights in the soothing effects of the boat's gliding as it responds to the smallest motion of the rudder. All this gives him the happiest notions of freedom: it affords him the chance to continue his scientific trains of thought in a daydreaming atmosphere: theoretical musings conjure up all kinds of images; there is no reality except in one's imagination. As his hands take the rudder, he will begin to concentrate intensely on his latest scientific ideas. So pervasive is the summer glare as it mixes the abstract meanderings of his mind with the deepest emotions of the scientist at work that there is an evident blending of a truly free life and a compulsive work ethic. He steers the boat with the agility and the lack of concern of a child. He does the hoisting of the sails himself, he clambers about to tighten knots, he maneuvers hooks and stakes to free the boat from its mooring. His face reflects his joy at such handiwork; his words and his smile show a contagious happiness.[2]

The Princeton physicist **Eugene Wigner** spent some time in Berlin in the early 1930s. Of his sailing adventures with Einstein, he wrote:

It was the greatest honor for the visitor to be invited on a sailboat outing. There was an intimacy that established itself on board between this extraordinary host and his friends or colleagues. Only mental demands were made—I do not recall anybody contributing practical work to on-board activities. Those, Einstein liked to see to by himself: providing for sails and rudder while at the same time filling his pipe and seeing to the exchange of ideas by all the others. And at times, he just sat there and dreamed. At such times, only water, wind, Sun and specks of conversation would be around—I could not have enjoyed it more.[3]

As the boat came to a stop at the edge of the shoal, Einstein broke the spell of silence: "This is the kind of sailing I used to like best: no

FIGURE 11.1 Albert Einstein in 1930, while sailing on Lake Templin. (*Photo:* Hermann Landshoff)

wind, no stress about doing the right thing as a sailor, just the reeds off to the sides. I could stay here for hours."

NEWTON: I'm afraid this is not the time and place for quiet thought, Professor Einstein. I'll remind you: you promised to tell me about your view of gravitation. I do have an inkling of what you are going to suggest: that we replace the spacetime plane in our two-dimensional example by a curved spacetime surface. Or, more generally, that we replace the four-dimensional world of spacetime by a curved analog. And I would guess that this distortion of spacetime, its "curvature," is the secret clue to gravitation. Am I right?

EINSTEIN: What else could it be? You put your thoughts in the right direction—and I owe you a compliment. I only wish I had had such a clear perception myself back in 1907; that would have saved me many a headache. But before we delve head over heels into the distortion and curvature of spacetime, I'll give you another simple argument: add to the old equivalence principle my theory of relativity, and together they will predict something that must appear almost sacrilegious to your ears. To wit, that gravity itself bends time. Or, to put it differently: gravity changes the flow of time.

NEWTON: Are you implying that the flow of time will change when we enter into a gravitational field? Let's be practical: We are presently in the gravitational field of the Earth. Now suppose someone were able to switch off gravity for a while. Do you mean to tell us that this would make our watches tick differently?

EINSTEIN: That's precisely what I am saying. I'll agree that this appears like a bit of a tall tale. But I am confident that I'll be able to convince you in due course; recall that my special theory isn't easy on the concept of time. You do remember time dilation. Indeed, were I able to switch off the Earth's gravitation for just a week, our clocks would accelerate a bit. The principle is straightforward: *the stronger the gravitational force in a given location, the slower the flow of time*. It is not that different from water: in a wide river bed, its flow may be slow, languid; but when it hits a narrow passage, its flow will accelerate.

HALLER: To keep matters even, let me note that this slowing of the flow of time has a relative implication only: this is not a fountain of youth for people who mind aging. In that respect, it is similar to time dilation. In any given place, the flow of time is universally prescribed. It changes only in relation to other regions of spacetime. On

Earth, for instance, the gravitational influence of both Earth and Sun make time flow the tiniest bit more slowly than somewhere in free space, far away from looming nearby masses.

NEWTON: And how great would that effect be?

HALLER: That we can figure out pretty easily. I'll describe a simple thought experiment, an experiment I perform only in my mind, not in a lab; this experiment is quite similar to one that Einstein used in 1907 in order to derive the effect of gravity on time dilation. Imagine we have a tower 300 meters high at our disposal; never mind this is not an experiment that is easily realized in reality—I repeat: this is an experiment we just imagine performing. At the top of the tower, we suspend a light source that shines light—say, a laser beam—straight downward. At the foot of the tower, let there be a receiver for the light emitted downward; this receiver will notice, accept, and evaluate the light signal. Let there also be an elevator inside the tower, pierced at the top and bottom so that the light beam can travel through the elevator without interference, Last, imagine there is an observer who has enough experimental tools to follow the light's progress.

The following situation now presents itself: the laser beam is made up of photons of a given frequency. We can measure this frequency. When an observer on the ground measures it, he obtains a measure of the flow of time. If another observer performs the same measurement at the top of the tower, his results should show the same numerical value for the frequency that was seen at its base. Should this observer come up with a different result, that would mean that the flow of time at the top of the tower differs from that at its base.

Let me add that the elevator, which we imagine is at the top of the tower as we start our experiment, does not constitute an inertial system; it is, after all, located in the Earth's gravitational field. That will change as we start our experiment: we drop the elevator down in its shaft, breaking its free fall just before its arrival at the bottom, to avoid a severe crash.

NEWTON: Let me imagine what will happen. As it starts its free fall, the elevator changes into an inertial system. This follows from Einstein's principle that inertia and gravitation are equivalent. I surmise that I am now able to observe the light signals the same as I would do were I an observer somewhere out in the universe, far away from gravitational forces; this is what the equivalence princi-ple demands.

HALLER: That's exactly what I was leading up to. By putting the elevator into free fall, we eliminate gravitation inside it. If we observe the light signals passing through the elevator, we see them propagate with the usual velocity of light. This is, after all, an iner-tial system.

NEWTON: Now wait a minute. The elevator we are observing is falling down, subject to steady acceleration. And you still mean to say the velocity of light inside it remains the same?

EINSTEIN: Never mind that the elevator, when seen from the tower, is moving downward in accelerated motion—this does not imply that light will also move in accelerated motion with respect to the elevator. The speed of light remains a constant. That is a must—just recall the equivalence principle. It says that a system in free motion is an inertial system; and in every inertial system, my special theory of relativity applies: the velocity of light is the same old c. You could say that the elevator and light beam jointly ignore gravitation. As far as they are concerned, gravitation does not exist.

HALLER: But here is my main argument: let's look at the beam of light just as the elevator starts its free-fall motion. At that instant, the elevator is still at rest with respect to the laser gun. Light detec-tors see the signal that enters the elevator at the top, and leaves it through the opening at its bottom, before the signal hits the receiver below. The frequency of the laser light is measured right at the instant of light emission—and it is the standard frequency of this laser. What will it be as it reaches the receiver at ground level, 300 meters below?

It takes light only a millionth of a second to traverse the 300-meter distance to ground level. During free fall, the elevator also moves downward, if not by all that much: it drops at almost 5×10^{-12} m, which is one-thousandth of a centimeter per second.

Now, what does an observer inside the elevator notice at that moment? From his vantage point, the receiver moves toward him at a velocity of 10^{-5} m/sec. But that implies a change in the frequency of the light as it hits the receiver. This is due to a phenomenon named after Christian Doppler. This phenomenon—the *Doppler effect*—is best noticeable when we deal with sound waves, the frequency of which changes as a source approaches or moves away from the lis-

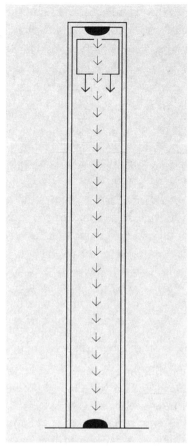

FIGURE 11.2 Thought experiment on the gravitational warping of time: inside a tower that is 300 m high and houses an elevator, a laser light source emits light vertically downward, traversing the elevator. An observer in the elevator watches the light beam on its way from its source to a receiver on the ground. As the elevator moves downward in free fall, the equivalence principle says it constitutes an inertial system. As the elevator starts falling, the observer measures the light frequency to be the same as when the elevator was at rest. As the light beam reaches the receiver on the ground, the elevator has moved down a bit, at a velocity which is still small. Due to the Doppler effect, the frequency of the light as it reaches the receiver has increased by a bit. As a consequence, the flow of time is different at the top of the tower than at its bottom.

tener. We all know that a police car approaches us with a siren of higher pitch than when it moves away from us, or even as it just stands there. It is easy to understand why: when the police car comes toward us, the time difference between two subsequent sound waves is smaller than if the siren is at a fixed distance; the difference is given by the ratio of velocities of the car and of the sound wave. That's why the sound has a higher pitch as the car approaches—and a lower pitch once it moves away from us. Now, the time interval between the two wave maxima is becoming greater. If we translate this phenomenon into what happens in our elevator experiment, the change in light frequency is given by the ratio of the velocity of the elevator and that of the light signal, i.e., 3.3 times 10^{-14}, or about three-thousandths of a billionth of a percent. That is a minuscule amount—but, as we will see, it *is* measurable.

NEWTON: For visible light, the frequency is correlated with the color; so we are dealing with a blue-shift here, that is, a shift to a higher frequency. I surmise the opposite will be true when the laser source is on the ground, while the receiver is at the top of the tower?

EINSTEIN: In this case, the receiver moves away from the observer in the falling elevator; and that means we are dealing with a *redshift* here.

HALLER: It turns out that there is a simple way to make the frequency shift plausible. The light that shines from above, we might say, "drops down" in the Earth's gravitational field, just as the elevator does. As they drop down, the particles of light, the photons, gain a little energy. Hence the blue-shift—because blue light is made up of higher-energy photons than red light. Similarly, the photons that are emitted upwards from the laser source on the ground will lose a little energy to gravitation—hence the redshift.

NEWTON: All right, gentlemen—but where is the connection to the behavior of time in spacetime?

EINSTEIN: It's staring you right in the face. Notice that there is little basic difference between light—which is nothing but a pulsating electromagnetic phenomenon—and a clock. A laser that sends out light of a precisely defined frequency can double as a very accurate clock. Imagine we compare the light emitted from the ground next to where our observer stands with that emitted by a similar laser mounted at the top of the tower. The light that hits the observer from above has a slightly enhanced frequency with respect to that emitted next to him. This implies that the time difference between the two light maxima is a bit longer for the laser on the ground than for the ray arriving from above. And, given that these time differences are a measure of the time elapsed, we might conclude that time runs more slowly on the ground than at the top of the tower. To be sure, the effect is minuscule. As we said, the difference amounts to three-thousandths of a billionth of a percent.

HALLER: If we really want to see an effect, we should engage in another experiment. Assume we have two precision clocks that advance synchronously while arrayed next to each other on the ground: the time they indicate is identical. Next, we move one of these clocks to the top of the tower and leave it there for a while. When we then take it down again and compare the times indicated by both clocks, what do you expect to find? We'll see that the clock that spent a while at the top of the tower now appears fast. Sure enough, the effect is tiny—but then, we are discussing the principle here.

EINSTEIN: So let's suppose a man spends eighty years at the top of the tower, then totters back to ground level—now he is older than the one who remained "grounded" all these years, by a whopping ten-thousandths of a second. Let's just note that our example here leads to a generally negligible and immeasurable effect.

HALLER: You'll be surprised, but this effect was actually verified experimentally. The first time this was done was at Harvard University, in 1960. But nowadays, experimental evidence for the distortion, or warping, of time is a routine affair.

EINSTEIN: Well, I'll be damned! How on Earth does one measure such minuscule time differences?

HALLER: Our Harvard colleagues chose Jefferson Tower, part of the Physics Building, as the actual tower for the experiment, although its height is just 22 meters. Also, they did not use laser light—they just took **gamma rays**—i.e., high-energy electromagnetic radiation. These were generated from unstable iron nuclei, which emit photons of a precisely defined frequency when they decay. To execute the experiment with all due care, they had to see to it that there wouldn't be any atomic motion to change the photon frequency. That they did by using a clever effect discovered about 1960 by the German physicist **Rudolf Mössbauer**—you'll forgive me if I cannot go into details here.

For the Harvard experiment, there was no attempt to measure the changed frequencies of gamma rays that hit the ground: rather, the atomic photon source was moved downward slightly, just enough to cause a slight redshift Doppler effect, that is, just enough to cancel the blue-shift caused by gravity. The end result was that there was no frequency shift at all at ground level, once the velocity for that movement was correctly chosen. Note that it was very small—about 2 mm per hour.

NEWTON: And to what precision does the result back up Einstein's prediction on the warping of time?

HALLER: The Harvard experiment was repeated with increased precision in 1965. The agreement between what was expected on account of the equivalence principle and what was measured was good to 1 percent.

EINSTEIN: That is really a wonderful result! Not that I expected anything else—still, it is great to see nature acting out what we have set in motion!

HALLER: Nevertheless, let me mention that this cannot be seen as a confirmation of the general theory of relativity; it just confirms one of its consequences once we accept that gravity "bends" time. But as to the precision with which this was done, experiments have meanwhile improved a lot: using rockets, one successful confirmation was executed in Virginia in the summer of 1976. A rocket carried an atomic clock to an altitude well above 10,000 km. During the two-hour flight, this clock was in constant wireless contact with a ground-based clock. Meticulous analysis of those data led to a more than satisfactory result. The agreement between Einstein's theoretical prediction and the experimental result was good to 70 parts-per-million, more than a hundredfold better than the Harvard experiment.

NEWTON: It certainly looks like there is no room for doubt. My compliments, Professor Einstein! There is no way to chip away at your ideas on the warping of time by gravity. Not that I feel competent to imagine all the consequences! There are, as we know, billions of galaxies in the universe, with billions of stars each. And all that matter acts gravitationally! That means there cannot be any way to talk of a uniform flow of time in the universe. In every location we might choose for observation—be it on Earth or close to Sirius or, who knows, in the Andromeda Galaxy—there will be a flow of time that differs from other areas. True, the effects we have quantified so far have been exceedingly small; but we know there are regions in space where gravity's effects are much stronger. We expect, then, that there will be a gravitational warping of time of a much stronger sort.

EINSTEIN: Didn't Saint Augustine write, a long time ago: "Time is like a stream of events; its flow is inexorable: things appear, only to be torn away?" How right he was! But he cannot have known that the flow of time varies from location to location. There are regions in our universe where time flows placidly, lazily; gravity keeps it from moving faster; there are others where time moves apace, unhindered. But there is no way of noticing those effects when we limit our observations to what concerns us immediately; after all, the flow of time in our own existence, in our lives, is somehow coupled to the external current of time. If the flow of time slowed up, so would our aging process.

NEWTON: Let's just suppose that there are regions in space where gravity acts more strongly than on Earth by several orders of magnitude. Then we might send a spaceship out there, leave it to roam around such a region for, say, a year, and call it back to Earth subsequently. Upon their return, the astronauts would notice that two years, or maybe two centuries, have passed on Earth, depending on the difference in gravitation they were exposed to. Gravity would act as a kind of time machine—is that imaginable?

EINSTEIN: Imaginable, yes. But we do not have objects at hand that might cause such violent differences in gravitational force. To change a year into a century, you would need one hell of a gravitational pull—and I have no idea where we might find it. Our planets, our stars, our galaxies won't do. I'll guess I have to be skeptical about it!

HALLER: For once, you are more conservative than our friend here, Mr. Einstein. Newton did not really get out of bounds there. We now have solid indications that there are objects in our universe that could warp time even more strongly. If our astronauts knew how to get in their vicinity, they should be able to bridge not only centuries, but as much as millions of years. But let's take it easy right now—let's revert to gravitation as we know it.

NEWTON: So be it. I do understand now that gravity can bend time out of its linear flow. What about the daily effects of gravitation? In our gedanken experiment, we had an elevator "drop" in free fall; that is what an apple does when it detaches from the branch where it grew. But why? So gravity warps the flow of time—but why does the apple fall to the ground?

EINSTEIN: That's also due to the bending of time, as you will soon see. But let me revert, for a minute, to the equivalence principle and the elevator in free fall: if I'm a passenger in that elevator as it falls, I have no notion of gravity. Instead, I'm inside an inertial system, just like the astronaut who travels across space. This says that our elevator is, as far as spacetime is concerned, a normal spacetime world such as we treat it in the general theory of relativity—there isn't a speck of curvature around.

NEWTON: Recall that we found that even in a curved space, the vicinity of any given point can be presented as a space without curvature, in terms of its tangential space. For a curved surface, this is the plane that touches it in this location. The system of the free-falling elevator, isn't that a kind of tangential space in spacetime, at some given point of spacetime?

EINSTEIN: That's precisely wh⌐t it is. Curved spacetime, like all curved spaces or surfaces, has the property that you don't notice its curvature in the immediate vicinity of any given point. That is why we can choose the relevant tangential space for a description of time and space—or better still, the tangential spacetime reference system. By that, we mean a system that propagates freely, i.e., in the absence of further forces, in spacetime, just like our free-falling elevator. The absence of forces is important for its being an inertial system. This implies a simple geometrical interpretation of the equivalence principle—we can safely eliminate gravitation because our curved spacetime permits us to use the relevant tangential space—or, better, the tangential spacetime.

When you glide downriver on a raft without rowing, you'll move as fast as the water around the raft. The "raft system" is like tangential spacetime, since the water current is nestled into spacetime. If, however, you start rowing, you move away from the surrounding current: your raft is no longer like tangential spacetime.

HALLER: The curvature of spacetime becomes apparent when we notice that the tangential space that we define at some event point is no longer able to describe the surrounding spacetime region. Here is a simple example: a skydiver falls freely before his parachute opens, if we pretend we can neglect the effects of air resistance. Due to the equivalence principle, he has no notion of gravity. But this is correct only because his dimensions are tiny when compared with the radius of the Earth. But let's suppose the skydiver is a 100-km giant falling toward the surface of the earth. He would be subject to the equivalence principle—or, more precisely, his center-of-gravity would be. During his fall his feet would be subject to a stronger attraction than his head, simply because the distance to the center of the Earth is different for the feet and for the head, and the gravitational force registers these differences accordingly over that distance. Now: during free fall, the average gravitational force is no longer noticeable, but there remain the tiny differences between that force's action on his head and his feet. As a consequence, the giant will experience a resulting force that tries to pull him apart. He feels his feet being pulled downward—they would like to fall faster than they can; his head is meanwhile being pulled upward—it would prefer falling a bit more slowly than the rest of the body forces it to do. This extra gravitational action on the giant's system is often called a gravitational tidal force—since there is a bit of an analogy between this idea and the tidal forces acting on the Earth's oceans. This effect is also a consequence of the curvature of spacetime: here, again, we see that only a small region around a specific point that we observe is being correctly described by the tangential coordinate system, which moves along. As soon as our observation includes larger dimensions, curvature becomes noticeable.

NEWTON: Do I understand correctly, then, that the decrease of the Earth's gravitational force with the distance between its center and that of an object on which this force acts, is a consequence of spacetime curvature? That the curvature decreases with the relative distance? That would be a hint that this curvature is closely connected with the warping of time we discussed before.

EINSTEIN: To make this more obvious, let's revert to our two-dimensional spacetime. It describes a distance between two of its event points that is uniquely given by the differences between their time and space coordinates. Let me choose, as two relevant points, the origin (which has $t = 0, x = 0$), and another point that I'll designate by $t = T, x = X$. The special theory of relativity gives the distance between the two points as $a^2 = (ct)^2 - x^2$ or simply the time coordinate squared, multiplied by c^2, minus the space coordinate squared. This distance is then fixed by the metric of the special theory of relativity—that is, by the metric tensor. Just as it happens with any normal surface, the transition to curved spacetime is described by a change in the metric tensor. Assume, for a minute, I define the distance in the vicinity of the origin by

$$a^2 = A(ct)^2 - Bx^2$$

where A and B are simply coefficients that will, in general, depend on time and location. Should that be the case, we know we are dealing with curved spacetime. The metric of the special theory of relativity presents a special case, with $A = B = 1$, hence its name "special." Here, spacetime is just a plane without any curvature. Still, it is not a plane with what we call a normal metric. Here, the metric of the special theory of relativity prevails—and let's recall it was not I who first introduced this metric; it was the mathematician Hermann Minkowski.

NEWTON: I'm getting it: the coefficient A tells us about the passage of time, while B gives us the spatial scale for the event point in question.

EINSTEIN: The bending, or warping, of time that we discussed before can now be easily understood: it is fixed by this coefficient A. There will be a time warp if the coefficient A does not remain constant as a function of time and/or location. In the case of our own Earth's gravitation, $A = 1$ for all practical purposes as long as great distances are involved, where that force field is negligible. But at closer and closer distances, A will eventually become smaller than 1. The closer we get, the greater the gravitational effect.

NEWTON: Now wait a minute! Let's consider a clock that is located at some arbitrary point in space, which we'll call X. Its world-line is given by $x = X$ and t, where t can take any value. Now the flow of its proper time t is given for our clock by the distance it has traversed on its world-line, a, divided by c, the speed of light. In the case of the special theory of relativity, we have $a^2 = (ct)^2$, which gives us the elapsed proper time as $t = (ct)/c = t$. And that means that our clock shows its proper time to be the same we would have expected from a clock at rest. If, on the other hand, A is smaller than 1, the situation changes. Now $t^2 = A(ct)^2/c^2$. That means the proper time that has elapsed is given by the time coordinate t, multiplied by the square root of A, where I remind you that A is dependent on the precise location, or x. Which implies that the flow of time depends on the location we choose to study—the flow of time is different when $x = 0$ than it is when $x = 100$ m. I surmise this is precisely the effect of the bending of time that we have been discussing.

HALLER: Just as you say. And it is important that the strength of time-warping depends on that of the prevailing gravitation. The smaller A is compared to 1, the larger the effect of gravity, as our example showed. We might therefore look at the coefficient A as a stand-in for the gravitational field that you introduced such a long time ago. For weak gravitational fields, such as those we are accustomed to from the Earth and the Sun, we can even give a mathematical expression for the relation between field strength and A— but we needn't worry about such details here.

NEWTON: Now let me get to the actual problem I see with gravity: I'll accept your notions on the curvature of spacetime and, specifically, on the warping of time. Our gedanken experiment with the free-falling elevator showed us that gravity impacts the flow of time directly. So far, so good. But tell me one thing: why on Earth does the elevator drop down in the first place? Why does an apple fall from the tree where it grew? Obviously, there must be a connection with the curvature of spacetime, but that still tells me nothing about the origin of gravity in the first place.

EINSTEIN: I'm happy you put the question so directly; but I would have come to that point in due course. What does the motion of some object in curved spacetime—say, that of the falling apple—actually mean? Let's take a step back to planar spacetime, the realm of the special theory of relativity. There, we had shown that the world-line of a free object has the property we call "cosmic laziness." If I know that the object in question is here today and there tomorrow, I can be sure that a clock that moves along with that body from here to there will show a maximal amount of proper time for the path chosen by Nature. This is what we call the geodesic line. Along any other line we might have chosen, the proper time would be less. We also know that this path is the straight line that joins starting point and finishing point.

NEWTON: I can just about imagine what you'll say next: the transition from planar spacetime to curved spacetime happens when we change the metric tensor, that's all. That doesn't touch our principle of cosmic laziness. It means that some object in curved spacetime will also follow a line when it is not subject to some field of force—that is, when it moves freely.

EINSTEIN: Of course, what else could you expect? Cosmic freedom leads us directly to cosmic laziness. A freely moving object behaves in spacetime like a swimmer who is too lazy to do any strokes but is happy to drift along with the prevailing current—in our case, the current of spacetime. The apple does not fall from the tree—rather, one could say that it permits itself to be moved down by not resisting its being swept away in the flow of spacetime.

NEWTON: Let's assume I consider the gravitational pull of our Earth as just a normal force like, say, the electrical force, and I want to interpret the dropping of the apple in terms of the special theory of relativity, that is to say, in planar spacetime. As long as the apple is still attached to the tree, and is at rest, its world-line is a straight line, a geodesic line in planar spacetime. Once the apple drops down, it will be accelerated—its world-line is no longer a geodesic line. That would require the apple to move from branch to ground in constant motion—clearly an unrealistic notion.

But now let me interpret the curvature of spacetime as the phenomenon we call gravitation: what does that entail? For one thing, the world-line that refers to the apple at rest as it hangs on the tree can no longer be a geodesic line in curved spacetime. There is a force that the branch exerts on the apple; it is just as strong as the gravitational pull downward that acts on the apple. The first of these is an actual force, the latter one isn't. The force the branch applies to the apple is used to keep it from following the geodetic line, as it would love to do. The apple now will not be swept along the prevailing current of spacetime. But as the apple is detached from the branch, it will fall down. With regard to curved spacetime, this means it can finally move freely. As it falls, no force acts on it—instead, it now floats along in the inexorable current of spacetime, along a geodesic line.

HALLER: You're right, the free-falling apple follows a geodesic line in curved spacetime. It will follow the world-line that will make its proper time a maximum for the process at hand. Now, what does our observation of the falling apple teach us about the structure of spacetime? For one thing, as we said, the world-line of the apple is a geodesic line, as for all objects in free fall. And when we know the geodesic lines of a surface, we know all about its curvature. This, by the way, is also true for a normal two-dimensional surface with curvature. If somebody gives us precise information about the relative distances between a number of points on the map of a mountainous region, we can use this information to reconstruct the mountains and valleys on the map—better, to determine the curvature on the map in front of us.

NEWTON: Let me first get back to the apple as it hangs on the tree: let's say it hangs at a height of 5 meters above ground level. In spacetime, its world-line is a straight line parallel to the time axis, crossing the x axis at $x = 5$ m. I now look at two events, one for the apple at $t = 0$ and, ten seconds later, at $t = 10$ s. These two events are connected by the world-line of the apple. Since it is subject to an external force, its world-line is not a geodetic one in spacetime. Now tell me: there surely must be a world-line between these two events that maximizes the proper time of the apple, that is, the relevant geodetic line. What is it like?

HALLER: Of course this geodesic line exists. Not only that, but it is easy to find; we need no fancy mathematics for that, just a little consideration of the physics involved. The geodesic line would have to be a world-line that connects the two events and also has the property that a watch that moves along it will maximize its proper time. At the same time, this line will describe the free motion of some object such as our apple. So, what then would be the free motion of an apple the world-line of which connects these two points?

NEWTON: That would have to be an apple that is not at rest, but that flies straight upward. At time $t = 0$, it will have to be located at $x = 5$ m, where we also expect to see it at time $t = 10$ s; in between these two fixed points, the apple will have to move along the geodetic line, i.e., it has to move in free flight. That we can easily arrange. At $t = 0$, we have to pitch the apple up such that it moves vertically upward for 5 seconds, that it then reaches its highest point and, subsequently, that it again falls down so that it regains its starting point after precisely 10 seconds.

The whole thing is easily calculated: the starting velocity at event point 1 has to be 49 meters per second. The apple will then fly upward for another 122 meters before turning around and falling back down. Its world-line is just a true parabola—and that would also be its geodesic line.

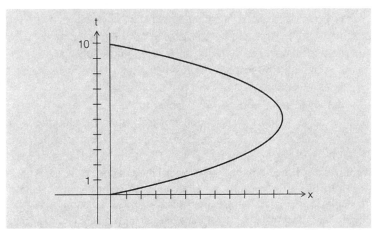

FIGURE 11.3 The world-line of an apple hanging on a tree (vertical line) and the geodesic line we obtain when we throw the apple upward such that it takes 10 seconds to fall back to its starting point (parabola).

EINSTEIN: That's where the time warp comes into the game: we already saw that time flows faster at higher altitudes. A clock that moves along the geodesic line, just as the apple that was thrown upward in our example, will be at a greater height than the apple at rest. When it arrives back at its starting point, it will indicate a time that is more advanced than that of a clock that has remained at rest. Sure, the effect is minuscule—but Nature isn't queasy about large or small effects—it is the principle that counts. And you will now agree that the principle of cosmic laziness is at work here on Earth.

It may sound strange, but it is nothing but the bending of time— that is, the change in the flow of time as a function of the elevation where it is being measured—that determines the free fall of an object. When the apple that fell from the tree finally hits the ground, the world-line that connects the event of detachment from the tree with that of hitting the ground is a geodesic line—it is the very line that joins them according to the principle of cosmic laziness.

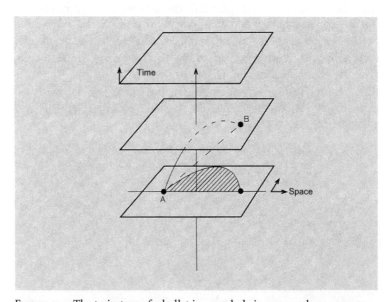

FIGURE 11.4 The trajectory of a bullet is a parabola in space as long as we neglect air friction (full shaded parabolic trajectory) from A to B. The world-line of the bullet is different: it joins A and B but is a geodetic line in the spacetime metric (dash-line parabola). The straight line joining A and B (dashed straight line) is not a geodesic line: a clock moving from A to B along this line will show a lesser time at its arrival at point B than will another clock upon arrival at B after traveling the geodetic line joining A to B.

NEWTON: Still, I am amazed that this tiny effect of bending time—an effect of the order of one-billionth in the case at hand—can actuate as drastic a happening as free fall. This fall, after all, changes the location of the apple by several meters per second!

HALLER: True enough, this is astonishing when we look at the order of magnitude. But we should remind ourselves that, while gravitation has changed the flow of time by just a very small amount, that this may well have larger effects on the locations involved: the scale factor that connects time and space is, after all, the speed of light. Multiply that billionth of a second by the speed of light, and we are dealing with 0.3 m. You see, we have no trouble reaching changes in location of the order of meters even though the time warp that causes these may be minuscule. The theory of relativity may well put time and space on one and the same level, but our customary units of meters and seconds simply are not suited for this context. Imagine you mention a given sum of dollars for one purpose, in terms of Italian lire for another. The actual value of the sum is not touched by the different currencies; it's just the numbers that differ by three orders of magnitude. The exchange rate between space and time is fixed by the velocity of light in the theory of relativity, and that means small time differences will appear on the books as large spatial separations.

NEWTON: If we now admit not just one spatial dimension into our scenario but two or even three, that should certainly not interfere with the principle of cosmic laziness. The world-line of a freely moving object—and that includes objects in free fall caused by gravity—will still be a geodesic line.

HALLER: Of course—the paths will still be geodesic lines in spacetime.

NEWTON: Let's consider throwing a rock: we pitch it upward at a given angle; it will follow a typical parabolic path. Would that count as a geodesic line in terms of Einstein's theory?

EINSTEIN: I can see your problem. You think a curved path such as the parabola of the upward-thrown rock could hardly qualify as a geodesic line. But I assure you it does. Recall that we are not really looking at the path of the rock here, but at its world-line—and that is a path in spacetime.

That is like projecting the path of an airplane on the surface of the Earth: the flight path is straight, but the path that its shadow follows

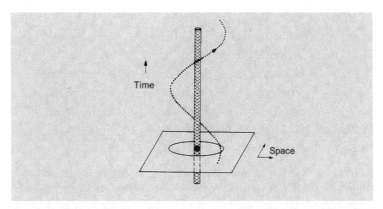

FIGURE 11.5 The world-line of a planet as it orbits the Sun is a helix. The plane in our sketch designates space (but only in two dimensions). The circular orbit of the planet in space is shown in the two-dimensional plane. This world-line is a geodesic line in spacetime.

on Earth must adapt itself to the structure of the surface. When the plane flies on a straight path above the Alps, the path of its shadow is anything but a straight line—it will jump back and forth all the time.

HALLER: This effect is even more impressive if we observe the motion of a planet around the Sun. Its orbit is, to a good approximation, circular. But in spacetime, it is a spiral, where the spiraling is directed along the axis of increasing time. But if you take the curvature of spacetime into account, such as that caused by the Sun in the case at hand, we can easily show that the path of the planet is indeed a geodesic line.

EINSTEIN: I can give you another example for a geodesic line that is by no means a straight one. Suppose the lifeguard at a beach hears a drowning person's scream for help. The lifeguard will not jump into the ocean immediately so as to reach the endangered swimmer as quickly as possible—he will run along the shore for a while before he dives in. By carefully choosing the place where he goes into the water, the lifeguard is able to minimize the time it will take him to reach the person who needs his help. In this sense, his path is a geodesic line, although it is not a straight line in space—there is an abrupt change in direction as he plunges into the sea. In our example, the geodesic line is defined by the path that makes time minimal. But in relativity theory we look for the path that maximizes time, following the principle of cosmic laziness.

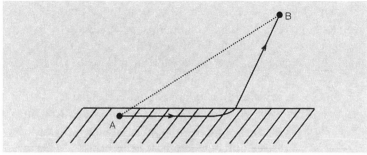

FIGURE 11.6 A drowning man at point B is being rescued by a lifeguard who approaches him from point A. The guard tries to minimize the time for reaching point B. Therefore, he does not take the geometrically shortest way (dotted line). Rather, he follows the corresponding geodesic line: this is the line that minimizes the time within which he will reach point B (i.e., the full line that bends upward in the sketch).

But enough of this laziness now. A breeze is starting up, and we are lucky it comes from the west; it will help us get home quickly. So let's stop our discussion for now and head to port.

Einstein steered the boat out of the reeds and set its course toward the northeast, heading straight for Caputh.

CHAPTER 12

Matter in Space and Time

> You will be convinced of the correctness of the general theory of relativity once you have studied it. That is why I will not defend it to you, not by a single word.
>
> —Albert Einstein (on a postcard
> to **Arnold Sommerfeld**, 1916)[1]

At dinner time, Einstein surprised his two colleagues with the idea to move the second part of the discussions to another place—to Pasadena, just east of Los Angeles. In the 1920s, Einstein had paid regular visits to Caltech, the California Institute of Technology. Caltech had, in the process, become something of a second home base to him. It was also at Caltech that Einstein had been made aware of the new cosmology that would become the most important application of his theory.

Haller, who had also worked at Caltech, and Newton agreed immediately to this plan. They decided to leave in two days. It turned out that Einstein had quietly made some of the necessary travel arrangements.

In the evening, Einstein finally spoke about the field equations of the general theory of relativity; they were rightfully seen as the crowning glory of his work on gravitation. Newton was full of anticipation.

EINSTEIN: So far, we have touched on a few aspects of my theory of gravitation, which I was able to develop starting from investigations into the equivalence principle of 1907; but this was before I developed the field equations on gravitation. Now the time has come that we can get to the core of the general theory of relativity; now we get to the equations that describe the dynamic interplay between spacetime and matter.

The fact that matter has inertia, that it likes to remain in a given state of motion, and that it will resist forces that want to change this state—that is the stamp which the geometry of spacetime imposes on matter. Shortly after setting up the special theory of relativity, it became intuitively clear to me that this couldn't be the end of the story. Somehow, matter should have a chance to take its revenge, to act back on spacetime; that way, we can ultimately speak of a real interaction between geometry and matter. In other words: matter bends spacetime, but spacetime uses inertia to fix the direction in which matter will move. Both impose their conditions. I asked myself how such a relation between matter and geometry might look. It should be as simple as possible; but it had to be complex enough to cover the multiple manifestations of gravitation that we know.

Also, it should be a local relation, so that the properties of matter at a given point of spacetime might fix the local geometric properties. In particular, it should describe curvature at a given location, just as we expect from mechanics. In a moving car, we know that there is a force acting at a given time and place that determines the acceleration—the force acts locally. Furthermore, the equations should establish the limitations of Newton's gravitational law.

In retrospect, it sounds quite simple, but—the number of detours I had to make between 1911 and 1915 before I hit on the correct equations was enormous; this I have to ruefully admit. The biggest problem was my lack of expertise in the field of non-Euclidean geometry. The importance of that aspect can be seen by the fact that even the great Göttingen mathematician David Hilbert also spent time in 1915 working on gravitation—but it took him only a few weeks to arrive at the correct equations by about the same time I did. However, he did have considerable difficulties with their interpretation in terms of physical phenomena.

HALLER: But it is correct to say that it is you, and not Hilbert, who managed to bring the general theory of relativity to completion.

EINSTEIN: Be that as it may—I certainly presented my theory and the field equations on November 25, 1915, at a meeting of the Prussian Academy of Sciences in Berlin.

First to the general concept: we had mentioned that the properties of curvature in any space or of any surface are not directly described by the metric tensor; rather, that is done by a mathematical construct we owe Riemann, and which is called the tensor of curvature. To honor Riemann, I might mention, one uses the symbol R to designate it. We don't have to discuss its precise form here. The important thing is that anybody disposed toward a few basic mathematical skills can determine the tensor of curvature from the metric tensor. R, by the way, is zero for any plane or for Euclidean space. The space we look at has to have some curvature before R will differ from zero.

NEWTON: Since the tensor of curvature describes any warping in a local way, I'll surmise that your equations establish a relation between R and matter. In other words, R should be equal to an expression that depends only on the matter.

EINSTEIN: You're right. I just wish I could have seen it that clearly in 1915. What exactly should be on the right side of the equation? In your old law of gravitation, mass is the source of gravitation; that means there should certainly be a term that contains mass. Otherwise there would not be an opportunity for the equations to contain your law of gravitation as a special case. But there must be more to it: we had seen that the equivalence principle makes sure that light, too, will be influenced by gravitation. The light particles do not have mass, just energy. That means the right-hand side of the equation must contain a term that covers the mass as well as the energy of the particles. And such a term does indeed exist. We call it the *matter tensor*, usually denoted by the symbol T. Sometimes we also call it the *energy-momentum tensor*, which expresses a physical quantity that has already played an important role in the special theory of relativity.

NEWTON: Of course, I know all about momentum and energy; but why do we need a matter tensor?

HALLER: The energy and the momentum of an object always relate to its full energy, its full momentum. For gravitation or for spacetime curvature, on the other hand, we are not interested in the total energy, but only in the energy density at a given location—that is, the energy per cubic centimeter at that location. The curvature of spacetime at a given event point is fixed by the structure of matter at this particular point, and not by its structure somewhere miles away. The curvature that is caused, say, by our Earth depends on the matter density—that means it also depends on the energy density in the Earth's interior. The matter tensor describes nothing more than the density of energy and momentum in a body: the total energy we find by adding up all individual energy densities or, mathematically speaking, by integrating them.

NEWTON: To put it briefly—the matter tensor is a quantity that describes the local energy-momentum properties of matter.

HALLER: We could express it that way. It is simple to write down the energy-momentum tensor for homogeneous physical systems—say, for a gas or a rigid body. We need not do that here; but we should at least mention that there is no trouble in writing it down.

NEWTON: So now we have the tensor of curvature to describe the geometric properties of spacetime, and the matter tensor to denote its energy-momentum properties. Now what about those Einstein equations?

EINSTEIN: You'll be surprised how simple they are. Let me first put them into words: the equations of the general theory of relativity state that *the tensor of curvature is proportional to the matter tensor*. Although we do not want to discuss mathematical equations here, I would nevertheless prefer to write down these equations; they are, after all, a symbol for the general theory of relativity, just like $E = mc^2$ stands as a symbol for the special theory:

$$R_{ik} = kT_{ik}$$

On the left side, you see Riemann's tensor of curvature R_{ik}; the subscripts i and k mean that this tensor has several components: i and k can be 0, 1, 2, or 3 (0 is the index for the time component, the other three indices stand for the space components). That means we have the components R_{00}, or R_{01}, R_{02}, and so on. The matter tensor is called T_{ik}, and just like the tensor of curvature, it has several components.

On the right-hand side of my equations, you also see a constant, which I call k. It is often erroneously called Einstein's gravitational constant: but, basically, it is nothing but the universal constant of gravitation which you, Newton, introduced a long time ago—to be precise, it is your gravitational constant G, but multiplied by $8\pi/c^4$. This constant determines the exchange rate between spacetime curvature on the left-hand side and matter on the right-hand side.

FIGURE 12.1 Albert Einstein in 1930, lecturing at Caltech on his gravitational field equations. Our picture shows him in the Athenaeum library as he discusses space devoid of matter, where the matter tensor vanishes, putting a zero on the right- hand side of his equation. (*Photo:* Hale Observatory, Pasadena)

HALLER: By the way, all quantities that have indices i, k, are symmetric: you can interchange i and k, and nothing will change. For example: $R_{01} = R_{10}$. The four indices 0, 1, 2, and 3 can be made into ten different pairs:

$$(0.0), (0.1), (0.2), (0.3), (1.1), (1.2), (1.3), (2.2), (2.3), (3.3)$$

This means that there is a total of ten Einstein equations—one equation for each pair of indices.

NEWTON: Well, these equations look simple and symmetric; but they certainly are no simplification when you compare them with my law of gravity, which is written in one simple equation—the more so when you take into account that both R and T stand for complicated expressions.

EINSTEIN: If you look at it only from an accountant's point of view, by counting equations, then you can say that my theory is ten times more complicated than yours. What is important is not the number of equations, but the conceptual simplicity. My equations are conceptually very simple, once you start from the idea of curved spacetime. Or do you have a better suggestion?

NEWTON: All right, let's leave it at that for right now. Your equations do state that the tensor of curvature on the left-hand side is nothing but the matter tensor multiplied by the gravitational constant. In other words: matter determines curvature.

EINSTEIN: That we can rightfully say, as long as you read the equations from right to left. But you can also read the equations from left to right. Then we would have to say that the curvature determines matter. Both statements are correct—geometry acts on matter, and matter reacts to geometry. Spacetime tells matter how to move, and matter tells spacetime how to curve. Space, time, and matter are united. This feature, however, makes my equations quite complicated—more so than any other set of basic equations in physics. They may look simple; but it is very hard to find solutions beyond some approximations. We will discuss one such solution in due course. But let's retain one thing: your theory of gravitation is a direct consequence of my equations, with certain approximations.

NEWTON: When I look at your equations, I would not get the idea that they hide my simple law of gravitation. Could you say that in a more detailed way?

EINSTEIN: Let's assume we look at a chunk of matter—say, an iron bar—and let that be at rest; then the matter tensor is, to a good approximation, nothing but the energy density at hand. However, in this case nine components of the matter tensor will vanish; and only one—the energy density—does not. This is the component T_{00}, the prevailing density of mass. We denote it by μ multiplied by the square of the speed of light, or μc^2. This is nothing more than my old expression $E = mc^2$ applied to mass density, to the mass in one cubic centimeter.

This lets my ten field equations reduce down to a single one—the one that has the (0.0) components. That is certainly progress. Now let's look more closely at the (0.0) component of the tensor of curvature on the left side. It turns out that this equation is almost the same as your gravity equation. This means that your theory of gravitation is a special case of my general theory of relativity. It relates to the case where all velocities of matter are much smaller than the speed of light.

HALLER: It is, we might notice, just this limiting case of the general theory of relativity which permits us to determine the constant k that Einstein's equations introduced. It is often called the Newtonian limit, and the "constant" that results is Newton's gravitational constant. You'll notice that the part played by the gravitational constant in Einstein's equation is much more general than what happens in your theory: there, this constant just gives the strength of mass attraction. In Einstein's theory, the gravitational constant gives the strength with which matter acts on the geometry of spacetime. If we could switch off this gravitational constant, there would be no such back-reaction: spacetime would not be warped, and there would be no such thing as gravitation. This means that the geometric structure of our world is closely connected with the gravitational constant; that should make it the most important "constant" in Nature.

Clearly, you can see that the gravitational constant is intimately connected with the curvature of spacetime. Multiply it by some given mass m—say, by the mass of the Earth—and divide the result by the square of the speed of light, or c^2, and you are left with Gm/c^2. This quantity is a length that can be measured in meters and centimeters. That means that the mass of one kilogram corresponds to a length of 0.74×10^{-27} meters. The mass equivalent of one meter, conversely, is 1.35×10^{27} kilograms.

EINSTEIN: You might then go to the bakery and ask for 10^{-27} meters of bread. If the baker had the requisite inkling of physics, he would give you 1.35 kilograms of bread.

HALLER: For completeness, here are the masses of the Earth and of the Sun, measured in meters:

Earth's mass: 6.0×10^{24} kg $\hat{}$ 4.44×10^{-3} m (that is, about 4 mm);

Sun's mass: 1.99×10^{30} kg divided by 1477 m (that is, 1.5 km).

In this way, we can convert any old mass into a length.

NEWTON: This conversion reminds me how we can convert space and time—express meters in terms of seconds by using the speed of light, which determines, if you wish, the exchange course of these two currencies: one second corresponds to 300,000,000 meters.

HALLER: This is completely analogous. In principle, the natural sciences and engineering could give up dealing with masses in terms of kilograms, and time units in terms of seconds, and simply describe everything in units of meters, just by using the speed of light and the gravitational constant. Unfortunately, the latter is not known precisely enough to make this a practical proposition.

NEWTON: Do we know today why the gravitational constant has just the value we measure—rather than another value ten times smaller, or ten times larger?

HALLER: I have to disappoint you there. To this day, nobody knows how to calculate the value of the gravitational constant. The origin of this important "constant" of Nature is unknown. It makes sense that there is some connection with the mass problem—after all, it is the mass of the elementary particles that make up matter—neutrons, protons, and so on. We expect to learn much more about the mass problem sometime in the near future, and that should also shed some light on the role of the gravitational constant.

EINSTEIN: I meant to discuss one particular solution to my equations that you might find interesting. But our sailboat excursion did tire me some. And we did, after all, look at the equations, and that means we took a decisive step.

There will be another day tomorrow. I assure you, Sir Isaac, that it will bring you some news—or better, some surprises.

CHAPTER 13

A Star Bends Space and Time

This idea has to be thought all the way through—it
is of a singular beauty; but way above it all, there is
this smile of inexorable Nature, cast all in marble
. . . the smile of Nature that instilled in us more
yearnings than intelligence.

—Albert Einstein[1]

The last day in Caputh had started. Early in the morning, Haller and Newton took a short stroll to Lake Templin. After their return, the meeting started on the terrace.

EINSTEIN: As I told you yesterday, I would like to talk to you today about one particularly interesting solution to my equations. When I introduced my general theory of relativity for the first time on November 25, 1915, at the Prussian Academy, all I could do was sketch approximate solutions for these equations—among them the border-line case of Newtonian mechanics, or even the gravitational time warp. It was not clear to me whether there is an exact solution, in closed form, for simple cases.

Karl Schwarzschild, director of the Astrophysical Observatory in Potsdam—not far from here—was among the first to try and address the mathematical properties of my equations. He was not in Potsdam at the time: the war was on, and he served in the German army at the Russian front.

Schwarzschild tried to find a solution for the equations in the simplest case we can imagine—for the case of a star with spherical symmetry and a mass density—or, more generally, for a sphere of a given mass. First, he tried to find the solution for the equations in the space outside the star—that is, in empty space.

NEWTON: Outside such a sphere, the matter tensor is zero. Your equations simply say in that case that the left-hand side, which includes the tensor of curvature, vanishes. A possible solution is

FIGURE 13.1 Karl
Schwarzschild (Emilio
Segré Archives, AIP)

surely empty Euclidean spacetime—that is, the space of the special theory of relativity without any curvature.

EINSTEIN: Sure, but Schwarzschild was looking for another solution where the left-hand side of the equations is zero as before, and which shows spherical symmetry. That means it has the property such that the tensor of curvature has the same structure in all directions in space; it depends only on the distance from the center of the sphere.

NEWTON: The result is very simple for my theory of gravitation—of course, the gravitational field of a sphere at rest is spherically symmetric and extends evenly in all directions. It depends only on the distance from the origin—that is, on the radius—and not on the direction. The gravitational force decreases with the square of the distance from the center of the sphere.

EINSTEIN: It is not that simple in the general theory of relativity. Still, Schwarzschild's solution is quite simple and has a few amazing properties. He calculated the metric tensor for the case we mentioned—that is, the metric structure of spacetime outside the star. The latter can depend only on the time and on the radius, i.e., on two of the four coordinates.

NEWTON: Wait a minute! Let me see how that would look. Since we are dealing with gravitation due to a sphere, I would expect mainly two effects. For one thing, there should be evidence for time's warping by gravitation. When I move a clock closer to the sphere, time will run more slowly. Next, gravitation should cause the curvature of space. I am anxious to see how that works out in detail.

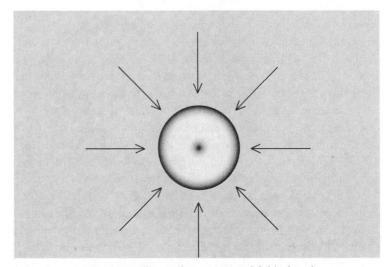

FIGURE 13.2 In Newtonian theory, the gravitational field of a sphere at rest—a star or a planet, for example—is spherically symmetric. No direction in space stands out. The gravitational force decreases with the square of the distance from the center of the sphere.

EINSTEIN: Since it is impossible to imagine a curved three-dimensional space, I use the following trick: I arrange a plane through the center of the sphere; then I study the metric properties of this plane, which would be just a Euclidean plane in the absence of any curvature of space.

I now look at circles around the origin. Their circumference is $2\pi R$ where the radius R is defined by the corresponding circle. When I am at a great distance from the sphere, the curvature is not noticeable. When I run from any given point on that surface in the direction of the sphere—say, from a point with radius R to another one with radius R minus 10 meters, I cover just 10 meters. When I run further such that the radius of the points I reach becomes smaller and smaller, this is no longer true. In order to reduce the radius by 10 meters, I now have to run 11 meters, then 12 meters, finally, 100 meters and more. This is due to the curvature.

It is like climbing a mountain. When we approach the mountain in the plane, the horizontal distance from the center of the mountain decreases by the path length we cover. But as soon as we begin to climb, the length of our path increases just as the radius decreases: recall that the radius is the distance between us and the

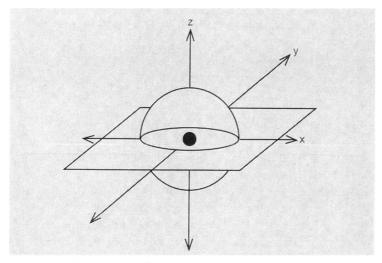

FIGURE 13.3 The curvature of space around a massive sphere can be illustrated by studying the metric properties of a surface through the center of the sphere. Our illustration puts that plane along the x and y axes.

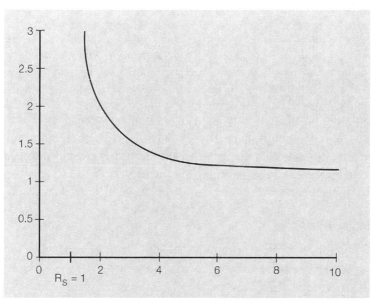

FIGURE 13.4 The squared ratio of the actually traversed path and the difference of radii in the case of the Schwarzschild solution, as we approach the sphere. This ratio increases as we approach this sphere.

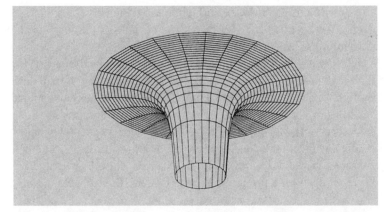

FIGURE 13.5 Metric structure of the equatorial plane around a massive sphere: its shape resembles a funnel.

projection of the summit onto the plane on which the mountain rests. The steeper our path, the stronger the effect. The ratio of the path covered to the difference in the radii—more precisely, the ratio of the squares of these quantities—is shown in figure 13.4.

The curvature of the observed surface is such that, when it is represented in three-dimensional space, it looks like a mountain that gets steeper as its height increases—like a funnel, if you wish.

NEWTON: It is strange that the relation between the actual path and the difference of the radius not only keeps increasing, but that at a certain point, it tends toward infinity.

EINSTEIN: That surprised me also, when I saw Schwarzschild's solution for the first time. But you are right: at some given radius, the ratio becomes infinite. This value for the radius now has a special name—we call it the *Schwarzschild radius*. It is given by the mass of the sphere and the gravitational constant:

$$R_s = 2GM/c^2$$

This radius is precisely two times the length we discussed when last we spoke about how to express mass in terms of length—that is, kilograms in terms of meters.

Since we just saw that the radius and the actual path toward the inside are very different because of the curvature, it makes sense to look at the circumference of the circle that corresponds to some

given radius, rather than looking at the radius itself. The advantage
is that the circumference is easily measurable—in principle, at
least—by applying a tape measure. In this method, we are not bothered by the curvature of our space. The Schwarzschild radius then
corresponds to a Schwarzschild circumference (U, in the following),
which I will denote as:

$$U_s = 2\pi R_s = 4\pi GM/c^2$$

All circles around the center of a star with a circumference that
equals the Schwarzschild circumference form a sphere about that
center; for reasons we will talk about later, this is what we call the
horizon.

If you now put the mass of the Sun into the equation, the result
is the Schwarzschild circumference of the Sun. It is about 19 kilometers—a tiny value in comparison with the actual circumference
of the Sun. In general, we get the Schwarzschild circumference of a
spherical object by multiplying the ratio of its mass to the solar mass
by 19 kilometers. Your own Schwarzschild circumference, if you
happened to be of spherical nature, would be about 10^{-24} meters.
That is about nine orders of magnitude smaller than the diameter of
an atomic nucleus. Let's look at a few numbers:

	Mass in Kilograms	Schwarzschild Circumference in Meters
Proton	1.67×10^{-27}	1.6×10^{-53}
Earth	6×10^{24}	5.6×10^{-2}
Sun	2×10^{30}	19×10^{3}
Galaxy	10^{40}	10^{14}

Since the solutions Schwarzschild found for my equations apply
only outside the sphere—in the case of the Earth, outside the
Earth—you can see that the infinity of the Schwarzschild circumference presents no problems. This circumference is tiny compared
to the circumference of the Earth, so that the infinity does not really
bother us. By the way, Schwarzschild found the solution for the
inside of a sphere a short time after he had found it for the outside.
The infinity we discussed before then no longer applies for the
Schwarzschild circumference.

NEWTON: You are right, there are no problems for the Earth and for the Sun. But suppose somebody started compressing the solar matter more and more so that, ultimately, the circumference of the Sun became smaller than its Schwarzschild circumference of 19 km. In principle, I can even imagine all that matter being compressed into a single point—what we would call a pointlike mass. This would not affect the gravitational field at large distances. But then, the aforementioned infinity would now become a reality, and that bothers me.

In the natural sciences, we have a principle that everything that is permitted by the basic laws of nature may actually exist. What interests me here most is the fact that the apparent infinity would present a new property of gravitation, outside of what my theory admits. I consider this a pretty dubious result.

EINSTEIN: Nevertheless, I can reassure you that this infinity does not exist in nature, because matter cannot be compressed far enough to make an object smaller than the Schwarzschild circumference. Not everything that is permitted by my theory is actually realized in nature. After all, there are other forces of nature, other natural phenomena, in addition to gravity.

HALLER: If I were you, Einstein, I would not be so sure. We will have to get back to that when we get to Pasadena, and I'm afraid Newton is correct. Let's first look at the other important effect of gravitation, the warping of time. In this connection, Schwarzschild's solution also weighs in with a few surprises.

NEWTON: All right, let's take a close look at a clock operating near our sphere. As long as the clock is far away, its functioning is essentially not influenced by gravitation. The closer I get to the sphere, the stronger the effects of gravitation: the ticking of the clock changes—time flows more slowly.

EINSTEIN: Schwarzschild discovered that the curvature of time can be described by a very simple formula, which I would like to present. If t stands for a small time difference that is indicated by a clock far away from the sphere, then the corresponding time difference τ at radius r is given by $(\tau/t)^2 = (r-R_s)/r$. The ratio τ/t looks like the illustration in figure 13.6.

NEWTON: If I approach the sphere, the flow of time slows up, as expected. Its dependence on the radius should be such that, at least to a good approximation, the gravitational force should be given by my old law.

EINSTEIN: You are right. At large distances, your law of gravity actually applies. But there will be deviations at smaller distances.

NEWTON: A moment please—there is a strange effect once the radius gets close to the Schwarzschild radius. The flow of time slows up more and more; but at $r = R_s$ it actually stops completely. That is absurd, isn't it? Time actually comes to a stop?

EINSTEIN: Schwarzschild's solution says, in fact, that the flow of time stops at the Schwarzschild radius, that is, at the horizon. Gravitation causes something that nobody else would be able to arrange: it stops the flow of time, it "freezes" it up, one might say. But as I said before, there is no object in nature that is smaller than its Schwarzschild radius. Our friend Haller appears to be of a different opinion. As far as I am concerned, I also consider this standstill of time to be absurd. Nonetheless, I surmise the Old Man Upstairs has found a way to deal with it.

HALLER: Still, let's quickly discuss the prospects of what would happen if this phenomenon did exist. The crew of a spaceship, in the vicinity of such an object, might spend a few hours in close proxim-

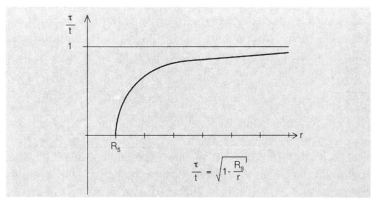

$$\frac{\tau}{t} = \sqrt{1 - \frac{R_s}{r}}$$

FIGURE 13.6 Change of the flow of time across the gravitational field of an object, as a function of radius r. At distances that are large in comparison to the Schwarzschild radius R_s, the ratio of proper time and the time far away from this object is barely smaller than 1. The closer we get to the Schwarzschild radius, the larger the difference. At a distance of 1.5 times the Schwarzschild radius, the flow of time has slowed down to 60 percent of its value at large distances. At the Schwarzschild radius, where $r = R_s$, the flow of time stops; it freezes up.

ity to the horizon; they will notice, once they get back to their home planet, that millions of years have passed.

EINSTEIN: That shows you how absurd the situation is. As far as I am concerned, I believe nature has devised a way to sidestep these curiosities at the horizon.

HALLER: Wait until we get to Pasadena. You'll soon have to revise your opinion.

NEWTON: You said earlier that there are deviations from my law of gravitation?

EINSTEIN: According to your law, the force of gravity decreases with distance—by the square of the distance. This does not apply precisely at small distances: in our own planetary system, it does not apply for orbits around the Sun with radii smaller than that of the Earth's orbit. The deviations are due, for one thing, to the influence of time's curvature in the Schwarzschild metric, but also to the warping of space. This is shown by the path of Mercury in our solar system: your theory would make that an ellipse; in the general theory of relativity this is only approximately true. It gives us an ellipse that rotates slowly around the Sun. Its orbit looks a bit like a rosetta with the Sun at its center.

When I discovered this consequence of my theory, I was of course aware that such an effect had long been discussed. Le Verrier discovered it in 1859. The planet Mercury describes an elliptical orbit about the Sun. But when you look closely, that orbit is not a stationary ellipse as expected from Newton's theory; rather, the ellipse changes constantly, such that the point closest to the Sun—called the perihelion of the ellipse—moves around the Sun slowly. The difference from a stationary ellipse is not large. Mercury's perihelion changes by a mere 43 seconds of arc per century. Notwithstanding the small size of the effect, the rotation of the perihelion of Mercury presents a serious problem for your theory: there was no explanation for this phenomenon.

Now, after I came up with my theory, there was a serious chance to solve this problem. Therefore, it was one of my first tasks to calculate Mercury's orbit. When the calculation was done, my result, indeed, was 43 seconds of arc. I could barely believe it.

To wit: I started from those abstract equations of mine, and without adding a single new parameter, I arrived at these 43 seconds of

arc. I had clearly hit a goldmine. There was no doubt, my theory had
to be correct. At the very least, it had survived the first serious test.

HALLER: With the aid of modern satellite technology, it is possible to survey planetary orbits more precisely than it used to be with
telescopes. Today's generally accepted value for the twist of Mercury's path is 43.11 seconds of arc plus or minus 0.21 seconds of arc
per century. Einstein's theory makes it 42.98 seconds of arc, and
within the margin of error, this gives us a brilliant agreement
between theory and experiment.

EINSTEIN: I would like to come back to the bending of light in
gravitational fields. If we start from the Schwarzschild metric of
spacetime, we can calculate the path of a light ray when it passes by
a massive object—say, a star—at a given distance. If there were no
gravitation due to this object, the path of the light would be a
straight line. Due to the curvature of spacetime, the path in spacetime is actually a geodesic line. In the framework of the prevailing
metric, this is a straight line—but not when we look at space only.
The light will be deflected by an angle which is given by the
Schwarzschild radius, divided by the shortest distance between the
light ray and the center of the object, multiplied by a factor of two.

We've already talked about the test of the prediction of the general theory of relativity that occurred during the solar eclipse of May
1919; as you may recall, it provided a confirmation of my theory.
Unfortunately, it also caused a big to-do with the press. I would love
to find out if there are new developments concerning this test of my
theory. At the time, there were rather substantial errors, and it was
not clear that my theory had been undeniably confirmed.

HALLER: The tests were repeated during later solar eclipses, and
the agreement between theory and observation was improved to an
accuracy of 10 percent. The most recent detailed observation was in
1973, in Mauretania. The deflection was determined to have a value
of 0.95 plus or minus 0.11 times the theoretical value.

In the 1960s, a new way was found to measure the deflection of
electromagnetic waves by the Sun; it was done with the help of radio
telescopes that are capable of giving us a very precise measurement
of very distant radio sources. These are at the centers of distant
galaxies. They emit strong radio-frequency radiation.

The advantage of this method lies in the fact that, here, we don't

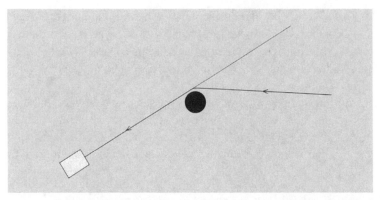

FIGURE 13.7 The gravitational force around a massive, spherically metric object changes the path of a light beam passing in its vicinity. The angle of deflection is $\delta\phi = 2R_s/\sigma$, where R_s is the Schwarzschild radius and σ is the shortest distance between the beam and the center of the object. In the case of the Sun, this amounts to 1.75 seconds of arc as the light beam barely touches the edge of the Sun.

have to wait for a solar eclipse. In the mid-1970s, detailed measurements were performed, reducing the error by a factor of 10. The agreement between Einstein's theory and observation is now at the level of about 1 percent. That means that the general theory of relativity passed this test with flying colors.

I would like to touch on a phenomenon that provides a direct consequence of the warping of time by gravitation. The change of the flow of time caused by the gravitational influence of a star, such as the Schwarzschild metric describes outside that star's matter: it is not very large in most cases—but it is not negligible. On the Sun's surface, for instance, time flows more slowly than at some distance from the Sun, by about two parts in a million. In one year, this adds up to about one minute. Of course, it is not possible to place a clock on the surface of the Sun. On the other hand, we do know how to use matter on the Sun's surface as a clock.

EINSTEIN: Are you talking about the spectral analysis of solar light?

HALLER: Yes—the atoms of the gases on the Sun's surface act like little oscillators that emit light of some given frequency, and therefore act like little clocks. Now, since time flows more slowly on the solar surface, the frequency of that light will also slow down by the same two parts in a million.

EINSTEIN: I wondered about that when I worked on the equivalence principle. At the time, I concluded that this effect is too small to be observable. Has that changed?

HALLER: Indeed it has. In the 1960s a group of Princeton physicists succeeded in measuring the effect; the result, not surprisingly, was in excellent agreement with the predictions of the general theory of relativity. A further discovery that concerned the deflection of light was made in 1979, bringing up a totally different aspect of gravitation. Measurements were made of what we call a double **quasar**—that is a system of two quasars in very close proximity to one another; they are separated by a very small angle. In the case we are talking about, an angle of 6 seconds of arc. Such small values actually occur quite frequently, so that there is no reason to expect any special features. But then it turned out that the two quasars were twins. Their physical properties, for instance—the velocity with which they move across the universe, and their atomic spectra—were identical. The only difference was that one of them emitted slightly weaker radiation. It did not take much time before the indications became obvious that we were dealing with two different images of one and the same object.

EINSTEIN: Well, that does sound like some kind of gravitational lens. It was about 1930 when I contemplated such a possibility. Also **Fritz Zwicky**, a Swiss astronomer working at Caltech, had similar thoughts.

NEWTON: And how should that work? How would we see two pictures of one and the same object?

EINSTEIN: That's quite simple. Suppose you observe some distant object—say, a distant star. It is entirely possible that its light passes close by some massive object on its way toward Earth—for instance, by a distant galaxy. This galaxy will cause deflection, thus providing a possibility that the light will reach Earth on two different paths— we might say that it passes by that galaxy once on its right-hand side, and once on its left. We might even think of more than two ways for the passage of this light, depending on the structure of the object in the light's path. Indeed, we do see two or more different images of the same object in individual cases.

HALLER: It has meanwhile been established that these predictions of Einstein's were correct. Gravitational lensing effects have proven to be observable in cases of very distant objects like quasars. Their

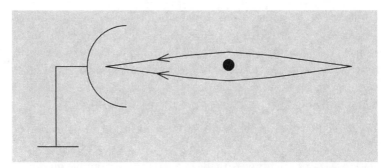

FIGURE 13.8 This figure shows the principle of a gravitational lens. The light, or electromagnetic radiation, that comes from a distant source is deflected by an intervening massive object, e.g., a galaxy. It can reach the Earth along two different paths, leading to two images of one and the same object. This effect was first observed in 1979 by astronomers in Arizona.

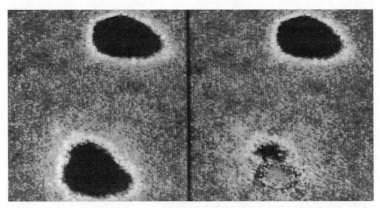

FIGURE 13.9 An example for the effect of gravitational lensing caused by a massive galaxy. Our picture shows two quasar images due to one and the same object. In the right-hand image, one of the quasars was electronically eliminated. In this way, we can also see the galaxy that causes the lensing effect. (*Photo:* University of Hawaii)

light is deflected by massive galaxies as the light travels on its way to Earth. It would actually be possible to calculate the lens effect if we knew the mass of the galaxy that disturbs the path of the light.

EINSTEIN: I don't think it makes sense to do more tests of my theory with such lens effects. Rather, we might do it the other way around. We know the theory, so let's consider the effect. We can then estimate the mass of the distant galaxy from the amount of deflection. In fact, I don't know of any other way to measure the mass of a galaxy. In principle, we could measure the mass of the Sun by observing light's deflection by the solar mass, according to my theory. But we know the mass of the Sun anyway—we don't have to rely on this method.

HALLER: You are quite right. The method of gravitational lensing today constitutes an important method of determining the masses of distant galaxies. The mass of our own galaxy, by the way, is not well known at this time. Anyway, we hope to obtain valuable information on the structure and on the mass distribution of galaxies, using this method. It makes up an important field of research, as we speak.

EINSTEIN: Gentlemen—time has passed quickly. We have come to the end of our stay in Caputh, my summer retreat here. I still have to make a few preparations for our departure. Good-bye for now.

CHAPTER 14

The Cemetery of Stars

Our Earth has been around for more than a billion
years. As to the end of her existence, I would advise:
let's just wait and see.

—Albert Einstein[1]

Adrian Haller arrived in Pasadena the next evening after stopovers
in Zurich and Chicago. He was well acquainted with the guest house
at Caltech, the Athenaeum, where he had stayed before. It is an old
classicist building on the east side of the campus, close to the line
that separates Pasadena from the residential town of San Marino.
The Athenaeum has served the same purpose ever since the found-
ing of Caltech in the 1920s. Einstein always stayed there during his
visits to Pasadena.

After getting settled in his room, Haller went to the reception
desk to inquire about Einstein and Newton. They had arrived in the
afternoon, but had retired to their rooms because of jet lag. A first
meeting in the new location was scheduled for the next morning.

Due to the time difference, Haller awoke rather early on the following day; he decided to take a short walk. The olive-tree-lined
main walkway that makes up the axis of the campus was deserted.
Haller was walking in the direction of the Millikan Library, the only
tall structure on campus, when he heard a familiar voice. Einstein
was sitting on a bench by the pond in front of the library, as he called
out: "Welcome to Pasadena, Haller. I hope you had a pleasant flight.
I assume you walk around campus for the same reason I do. The
time difference of nine hours is hard to take. When I was here sixty
years ago, this problem did not exist. It took two weeks by boat to
travel this distance; that was enough for a smooth transition into the
new time zone. While we are able to stretch or bend time with the
help of the theory of relativity, we don't have any way to displace it.

FIGURE 14.1 Porch structure of the Athenaeum, the guest house of the California Institute of Technology in Pasadena.

"Yesterday afternoon, Newton and I made the rounds of the Caltech campus. The Athenaeum is still as it used to be, as I remember it from the 1920s. Apart from that, the campus is hardly recognizable. There are new buildings all over—just the old Kellogg and Sloan Laboratories look the same as ever."

HALLER: When you were here, Caltech was in its first stages of development. After World War II, it advanced to become the most important educational institution in the area of science and technology for all of the American West. A major part of the experimental and theoretical research in the areas of astronomy and astrophysics happened here—not least among these the confirmation of your theory of gravitation. That happened either at Caltech proper or at the observatories on Mount Wilson and on Mount Palomar, both connected with Caltech. There were other contributions from satellites that were guided from the Jet Propulsion Laboratory, north of here. This will be, I assume, our main topic over the next few days.

EINSTEIN: I can see that, entering into these new territories for discussion, Newton and I must count as novices while you will be the leader of the expedition. But shouldn't we start with a good American breakfast?

After sharing breakfast in the dignified old refectory of the Athenaeum, our three physicists retired to a small meeting room in the north wing of the building, ready to resume their discussions in this scientific outpost of the New World.

EINSTEIN: Haller, let me get back to a remark you recently made concerning the Schwarzschild solution of my equation. I was of the opinion that the circumference of a celestial body could never be smaller than its Schwarzschild circumference because of the enormous matter density that would entail. You did not agree with that. Your protest reminded me of Fritz Zwicky, the Caltech astronomer to whom I often talked during my visits here. He was famous for having many speculative ideas, most of which turned out to be nonsense. For that reason I did not take him too seriously—but then, sometimes his ideas did have a speck of truth hidden in them.

HALLER: Let me give you a brief summary of the subsequent developments. Fritz Zwicky was indeed the first to consider the possibility that there might be objects in the universe that have a matter density significantly greater than that of a normal star. Decisive for this idea was the discovery by nuclear physicists who, in 1932, found that the atomic nucleus is made up not only of positively charged particles—protons—but also of neutral particles which we call neutrons. Now, there are very strong forces that act between neutrons and protons. These forces bind the atomic nuclei together as tightly as we know it. Free neutrons are not stable; in a·very short time they decay into protons plus electrons and neutrinos. Neutrinos are electrically neutral particles that are related to the electrons; both belong to a group of particles called **leptons.**

NEWTON: Since neutrons decay into protons, electrons, and neutrinos, couldn't we say they are made up out of these particles?

HALLER: No, in particle physics it is an everyday process that such

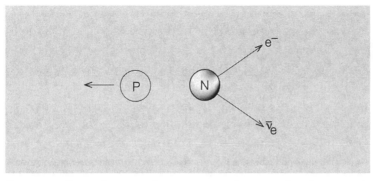

FIGURE 14.2 Decay of a neutron into a proton, an electron, and an electron-anti-neutrino. This is an example of beta decay.

transitions occur. But not every imaginable process is permitted. These transitions usually happen between particles that are some-what related to each other. That is the case between protons and neutrons: both consist of quarks, elementary building blocks that make up the particles we call **nucleons** which, in turn, make up atomic nuclei. The decay of a neutron into a proton is accompanied by the emission of an electron and a neutrino. And, in principle, it is nothing but the conversion of a neutron into a proton. But since its mass is about 0.1 percent larger than that of a proton, the energy that is being freed—according to the equivalence of energy and mass—goes to the electron and neutrino that are emitted in the process. (Actually, the particle emitted with the electron is an antineu-trino—but that need not bother us here.)

NEWTON: Is it possible that the reverse reaction can also happen, that a proton changes into a neutron?

HALLER: This cannot happen with a free proton because of the conservation of energy; recall, its mass is smaller than that of the neutron. However, this process can in fact happen inside an atomic nucleus: inside a nucleus, the masses of nuclear particles are a bit different from those of the masses of the same particles when they are free.

The conversion of protons into neutrons is an important process in the physics of stars. You can see this when you consider the fol-lowing gedanken experiment: Let's look at gaseous hydrogen that is contained in a given volume. (I will assume that this gas can be arbi-trarily compressed.) For starters, everything is normal: hydrogen atoms, which are made up of one proton and one electron each, are in a gaseous state. We now raise the pressure. For a long time, nothing will happen beyond the fact that the atoms move together ever more closely; the hydrogen will liquefy, but we continue to compress it. Finally, we will reach a state where the atoms are so closely packed that their density amounts to about $10^{11}g/cm^2$, which would be equivalent to an incredible one million tons per cubic centimeter. Obviously, we cannot reach such a density in the laboratory—but our calculations tell us that if we continued the compression process beyond this point, this would start the reaction we just considered: electrons and protons would fuse together into neutrons, emitting neutrinos in the process.

EINSTEIN: Each reaction is: a proton plus an electron will change into a neutron plus a neutrino. Our hydrogen gas has changed into some kind of neutron gas.

HALLER: Basically, we are now dealing with one gigantic atomic nucleus made up entirely of neutrons; it has the unbelievable density of about $5 \times 10^{12}g/cm^3$. Let's call this state of matter a neutron gas.

NEWTON: Since this is a gedanken experiment, we can continue to compress. What happens next?

HALLER: Having gotten as far as this neutron gas, we also have arrived at the end of what we surely know today. But we may well assume that, upon a further increase in pressure, the neutrons will eventually fuse together.

NEWTON: Since neutrons are made up of quarks, doesn't that mean that we will finally be left with a gas made up entirely of quarks?

HALLER: So we might assume. That is why experiments are now being conducted in particle physics laboratories where atomic nuclei—as, specifically, the nuclei of lead—are being collided at high energies. The goal of these experiments is to prove that, in these collisions, such a quark gas might be generated, if only for a brief amount of time. That is what the calculations of particle physicists predict. At the end of our gedanken experiment, we are then dealing with an enormously densely packed assemblage of quarks—a sort of quark matter.

EINSTEIN: Up to this point, we have talked about theory and gedanken experiments. But I would guess that you did not tell us all that just to amuse yourself—you probably want to apply this to the physics of stars, don't you?

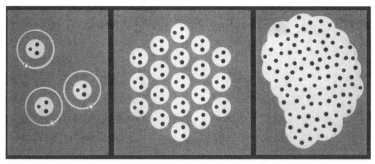

FIGURE 14.3 At high densities, a gas of hydrogen atoms is transformed into neutron matter. The assumption is that, as the density increases, it changes into a very dense quark gas.

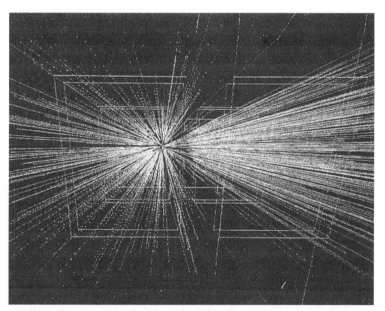

FIGURE 14.4 When the nuceli of lead collide at high energies, hundreds of elementary particles are emitted. One hopes that the analyses of such collisions will verify the assumption that, during the collision, a quark gas forms for a brief time.

HALLER: Yes, that is what we are getting to right now. For this purpose, it is best to consider the fate of individual stars. Stars, as we know today, get their energy from the process of **nuclear fusion** that happens in their interior. This is the same process that occurs in the explosion of a hydrogen bomb. A star, such as our Sun, is really a gigantic fusion reactor. Since we have now developed a pretty good understanding of what happens in the interior of stars, and since we can calculate this quite well with our computers, we can actually predict the future of each and every star.

Leaving out a lot of details, we can say that stars are relatively simple systems. Their future depends mostly on one quantity—on their mass. The reason for that is easily understood, as the example of our own Sun shows. It holds together simply because of its own gravitational pull. Were we to switch off this gravitation, the Sun would explode immediately just like a gigantic hydrogen bomb. That is, in fact, what happens all the time—but the explosion is made impossible by the all-encompassing gravitational pull.

In the center of the Sun, we are dealing with a delicate dynamical equilibrium. Solar matter would like to explode due to the nuclear process, but gravity wants to have the opposite effect, leading to implosion. And the result is what we see every day: our Sun in all its stability and durability, which has been around for more than four billion years.

The situation reminds us of the equilibrium in a hot-air balloon. The hot air originating from below inflates the balloon, while the pressure of outside cold air tries to compress it. Both forces are normally in balance, but only as long as hot air continues to be added from below. In a star, the hot air is replaced by the radiation that is generated in its interior; the cold air pressure from the outside is replaced by the gravitational pull of the matter in the star. Once the balloon runs out of fuel, it will fall to the ground and collapse. The amount of fuel determines the length of the balloon's flight.

This is analogous to what happens with our Sun. A few billion years from now, the Sun's nuclear fuel—that is, its hydrogen—will have been used up. The nuclear furnace will cool down, and the Sun will approach its death. The nuclear process will gradually die down, and solar matter will compress more and more, because of the prevalence of the gravitational pull. The Sun will turn into a brilliant **white dwarf** about the size of the Earth. Eventually, its radiation will decrease: its light will fade. The end is not all that exciting; what remains is what we call a black dwarf—a densely packed sphere made up of ashes.

The dynamical equilibrium of the Sun during its active phase of radiation is very much dependent on the solar mass, because that mass determines the strength of its gravitational pull. The larger the mass, the stronger that pull. Therefore, very massive stars have a higher temperature in the interior which will produce sufficiently large pressure for the radiation that wants to penetrate to the outside. That is why very massive stars are subject to the same penalty as very heavy humans—their lives are shorter than those of their "thinner" or less massive colleagues.

Stars that are lighter, or even a bit heavier, than our Sun will be subject to a fate similar to that of the Sun: they will pass through a phase as white dwarfs, finally to end up as black dwarfs. But note: the mass cannot be arbitrarily large. The Indian-born U.S. astrophysicist **Subrahmanyan Chandrasekhar** did a calculation in the 1930s, with the following result: normal atomic matter in a white dwarf can resist the pull of gravitation only if its mass is not larger than 1.4 solar masses. In other words, white and black dwarfs with masses larger than 1.4 solar masses cannot exist. The atoms in their center would not be able to hold the pressure.

NEWTON: But aren't there a lot of stars with larger mass? I am thinking of Sirius, which has about 2.5 solar masses. What will happen to them once they cool down?

HALLER: There is only one solution—what will happen is the same thing we considered in our gedanken experiment with hydrogen. Gravity produces pressure which grows stronger with time and finally demolishes the atomic matter. A star made up of neutrons alone develops—it is basically a huge atomic nucleus which, however, is tiny in comparison with the original star. A **neutron star** with about 1.5 solar masses has a radius on the order of about 10 kilometers.

The first theoretical investigations on neutron stars were done in the 1930s by **J. Robert Oppenheimer** and his colleagues in the United States, and by Lev Landau in the Soviet Union. Oppenheimer later on became the head of the Manhattan Project, which built the first atomic bomb. Landau was subsequently one of the leaders in the development of Soviet nuclear weapons.

NEWTON: So, what is the answer? Do we have any evidence that there is something like a neutron star?

HALLER: There is no direct proof, obviously: neutron stars do not send out signals that clearly point to the existence of neutron matter. But we have to look for indirect indications. If, for instance, we find an object in space that has a large mass and a tiny radius—say, two solar masses and a radius of 15 kilometers—then the object would have to be a neutron star.

In 1967, radio astronomers discovered the first pulsar. Pulsars are objects in space that send out electromagnetic signals in short sequences; these signals may be radio waves, optical light waves, or even X-rays. There are pulsars that give out hundreds of signals per second, sufficiently regular so that we could use them as clocks. These signals are, indeed, so evenly spaced that there were speculations that they were sent out by some distant civilization.

Today we know the reason for this regularity: it is the rotation of the pulsars. The mechanism is similar to what happens in a lighthouse: its rotating searchlight is the cause for the regular appearance of its beam. From the details of the observed signals, we can conclude that the radius of the rotating pulsar must be of the order of about 10 kilometers—about as much as we expect for a neutron star. This is a convincing, if indirect, proof for the existence of neutron stars. Not every neutron star is a pulsar, but conversely, every pulsar is a neutron star. Fritz Zwicky was the first to predict the existence of neutron stars in the 1930s. Nobody gave him any credence, but he was correct.

EINSTEIN: We saw that there is an upper limit for white dwarfs. How about a limit for neutron stars? There are many stars with masses that are large in comparison with that of the Sun. When they cool down, their gravitational pull is so enormous that it is not certain that neutron matter is up to the gravitational pressure. That means we might have an upper limit for the mass of neutron stars. If that is indeed so, we have an interesting question: what will happen when this mass limit is exceeded? Will that lead to quark stars?

HALLER: In order to find that out, we need to know a lot of details about the physics of neutrons and about the forces prevailing in atomic nuclei. The relevant experimental investigations were conducted in the 1940s and 1950s, largely in connection with the development of atomic and hydrogen bombs. Here again, we can see the ambivalent and universal character of the sciences. We can learn a

FIGURE 14.5 Albert Einstein on a visit to Caltech in 1931. *Front row:* Einstein
with Caltech president Robert Millikan to his right and the theoretical astrophysicist Richard Tolman to his left. Fritz Zwicky is the second to Millikan's
right. Both Zwicky and Tolman were key contributors to the formation of modern stellar models. (*Photo:* California Institute of Technology, Pasadena)

lot from the stars about atoms and their nuclei, and visa versa. The
equations that describe the collapse of a star into a neutron star are
also valid for the explosion of a hydrogen bomb.

When Oppenheimer started to calculate the formation of neutron stars, he made strongly simplifying assumptions. For instance,
he neglected the counterpressure of the imploding matter as well as
the role played by electromagnetic radiation—in particular by X-
rays. He concluded that there must be an upper limit for the mass of
a neutron star.

Other physicists assumed that the counterpressure of neutron
matter is so enormous that it can certainly resist the gravitational
pull, no matter how large that might be; that means the mass of neutron stars could be arbitrarily large.

The clarification of this problem was reached in the 1950s, after
many details of the properties of nuclear matter had been learned,
not least by way of the experiments and the test explosions that had
been implemented with nuclear weapons. So it was not by chance
that the same physicists who worked on the theoretical foundations
of nuclear weapons found the actual solution. In the United States it
was **John Wheeler**; in the Soviet Union, it was **Yakov Zeldovich**.

FIGURE 14.6 This photo shows John Archibald Wheeler around 1954. (*Photo:* Blackstone-Shelburne, New York, with permission of J. A. Wheeler)

To this day, we cannot determine the upper limit for the mass of a neutron star with the same precision that we have for the upper limit of the mass of a white dwarf; but it can be shown that there is an upper limit and that this limit is relatively low—just slightly more than two solar masses.

EINSTEIN: Hmm—but what happens when the mass is larger?

HALLER: Then the star will inevitably collapse, and it will bypass the formation of a neutron star. The neutrons in the center of the star will fuse and will form a nucleus consisting of quarks alone. But that doesn't give us any relief. We know about as much about the action of quarks as we know about what happens with neutrons at high pressure, but what we do know is due to the recent experimentation of our colleagues in particle physics. Quarks are certainly not capable of resisting the all-penetrating force of gravitation. The star collapses further, and there will be an inescapable catastrophe, which we call gravitational collapse.

EINSTEIN: I would never have expected that. That means . . .

NEWTON: Yes, Professor Einstein, that means that gravity defeats matter. The Schwarzschild solution to your equations is coming into play well beyond what you originally conjectured. The equations on gravitation that you formulated are now running the show.

All right, Haller. I can imagine what comes next. But right now, after this shock treatment, let's give our friend a brief break.

CHAPTER 15

The Wall of Frozen Time

For the likes of us who have faith in physics, the separation of past, present, and future has no more meaning than that of a tenacious illusion.

—Albert Einstein[1]

Haller suggested that they spend the second half of the morning in the park of the Huntington Library, an easy ten-minute walk. Einstein, who had taken frequent strolls around the grounds, agreed right away. Newton joined them, if a bit unhappily. The discussion about the collapse of heavy stars finally resumed on a bench below a tall stone pine in Huntington Park.

EINSTEIN: The cosmos is full of stars that have masses much larger than that of the Sun—by a factor of ten, twenty, or even more than that. If what you said before is correct—that neutron stars cannot be much heavier than two solar masses—then I ask myself what exactly will happen when a star like this reaches the end of its active life and collapses; as we saw before, that must lead to catostrophe.

HALLER: In the late 1930s, Oppenheimer examined this problem here at Caltech. He was an optimist and assumed that, in fact, there is no mechanism that can stop the collapse. Ironically, tests with the hydrogen bomb in the 1950s confirmed his assumptions, which had been considered to be too optimistic by a number of his colleagues.

EINSTEIN: Let's look at a star about ten times as large as our Sun. Suppose it has reached the end of its radiative phase. That means it is about to collapse. What then does happen?

HALLER: Let us make an additional assumption—let the star be of perfect spherical symmetry. Viewed from its center, all directions are equivalent. The spacetime metric outside the star is described by the Schwarzschild solution to Einstein's equations. On the surface of the star, the flow of time is slower than in the spatial regions far away. The collapse begins—the star's circumference gets smaller.

NEWTON: Wait a minute, how do you want to determine the metric outside the star? If the distribution of matter is static—that is, at rest—the Schwarzschild solution applies. This is no longer true as soon as the collapse begins. I would expect that the metric properties of spacetime outside the star will change with time, so that Schwarzschild's solution is no longer useful.

EINSTEIN: For once, you are wrong, Sir Isaac. Of course, the question poses itself whether the Schwarzschild solution, which was derived for the case of a distribution of static matter, will still prevail. The answer is quite simple. The American mathematician George Birkhoff proved in the 1920s that this is the case as long as we have spherical symmetry.

NEWTON: Excellent—the star can increase or decrease its radius, and the Schwarzschild metric will not change. That simplifies matters enormously.

EINSTEIN: The main thing is that the spherical symmetry remains intact. As it collapses, the star is not supposed to engage in escapades that make it crumble more quickly in one direction than in another. The Schwarzschild solution is valid even for a pulsating star, provided we always look at points in space outside the star. There is no way to gain any information on the pulsation of matter from the spacetime metric outside the star.

Still, it should be mentioned that the collapse of a star never happens without abandoning strict spherical symmetry. There will always be little deviations from it, but they don't change the basic process. This should not be astonishing to you. A similar theorem is valid in your theory of gravitation. The force of gravity that issues from a pulsating star does not change with time in your theory, Sir Isaac, as long as the spherical symmetry prevails. There is really no way to notice the pulsation from gravitational effects on the outside.

HALLER: Indeed, Birkhoff's theorem is a great help. Without it, a simple solution would not be possible.

Let's get back to our problem and follow the collapse of our star. Oppenheimer and his colleagues were the first to consider this in the late 1930s. Let us assume we are in a space station in circular orbit around the star at a distance of, say, fifty times the radius of the star, and that we are looking at its collapse. It starts slowly, but subsequently the diameter of the star decreases ever more rapidly. Stellar matter collapses. After a short time—to be measured not in years, but in hours—the circumference of the star is down to ten times the Schwarzschild value, then it decreases further to just four times this value.

By the way, the Schwarzschild radius is, for our example of a star with ten solar masses, about ten times as large as the Schwarzschild radius of the Sun, and amounts to 30 kilometers. The Schwarzschild circumference is then 188 km. If the star's circumference is four times the Schwarzschild circumference, the flow of time on the surface is 15 percent slower than far away—in our space station. This can be directly observed by a spectral analysis of the light from the star: the latter should have a redshift of about 15 percent.

A little later, the circumference of the star is only twice the Schwarzschild circumference. The flow of time on the surface is now 41 percent slower than at a greater distance from the star. From our point of view as we observe the collapse from a safe distance, this collapse slows up dramatically due to the change of the flow of time on account of the increasing gravitational force at the surface of the star, which approaches the horizon inexorably.

By way of illustration, let us imagine that, shortly before this collapse, a spaceship carrying an astronaut was placed somewhere not too far from the surface of the star; it has to keep its engine running constantly, to keep it from falling down to the stellar surface, due to gravitation. As the circumference of the star decreases during its collapse, the spaceship should remain at the same altitude by changing the thrust of its engine. How exactly this should be arranged does not concern us here; after all, this is just a gedanken experiment.

EINSTEIN: I should hope so. If this were a practical case, the astronaut would have to be suicidal.

HALLER: That is for certain. His prospects for the future would not be so hot, as we shall see. Be that as it may, let us imagine there is a very precise clock, both in the spaceship and in the observer's station. Let them send radio signals at regular intervals, to be received both in the space station and in the spaceship. As we start the experiment, both clocks advance synchronously. This changes quickly due to the gravitational time warp, as soon as the collapse gets going.

EINSTEIN: Should there be a difference in the way the clocks advance, it would be given by some time scale which presumably is related to the stellar mass.

HALLER: The characteristic time scale that now becomes important is indeed proportional to the mass of the star. It is simply the time it takes for light to cross a distance as long as the Schwarzschild radius. Let's call it the Schwarzschild time or, for short, T_s. In the case of the Sun, this time is 10^{-5} seconds; in our example, it is 10^{-4} seconds, or one ten thousandth of a second.

Now let's assume the circumference is twice the Schwarzschild circumference; the time warp becomes noticeable. It amounts to 41 percent. Now the surface of the star approaches ever more slowly the surface of the horizon; the time scale that gives the rhythm of the collapse is the Schwarzschild time. The difference between the star's circumference and the Schwarzschild circumference will decrease by a factor of 2 every 0.7 T_s, or 0.00007 seconds. After 0.00007 seconds, it is only half of what it was before; after a further interval of the same duration, a quarter, then an eighth—the horizon is being approached exponentially.

After only one thousandth of a second, the difference has already decreased by a factor of 20,000. The approach of the horizon is very rapid; but it will never be fully reached. Even a year later, the stellar circumference is still a tiny bit larger than the Schwarzschild circumference. Mathematicians call this an asymptotic approach.

Now let's get to the time warp. Since we can follow the advance of the clock in the spaceship, we have an easy time observing the gravitational change of the flow of time. When the stellar circumference is twice the Schwarzschild circumference, we measure a flow of time in the space station that has been slowed down by 41 percent. Then everything happens very quickly. The time warp increases rapidly, in an exponential way. Every time 1.4 T_s, or 0.00014 seconds, have elapsed, the flow of time we see in the spaceship decreases by a factor of 2. After the passing of only one thousandth of a second, the current of time has already been reduced by a factor of 141. This implies that the frequencies of the incident light decrease quickly— blue light changes to red light, then to infrared, finally to radio frequency; X-rays change to visible light, visible light to radio waves.

EINSTEIN: Since the frequency increase happens rapidly, this means that the darkening that happens as the star "freezes up" occurs very rapidly—in a fraction of a second—once the circumference of that star comes close to the Schwarzschild circumference—or, we might say, as we reach the horizon. Correspondingly, the collapse of solar matter by gravitation is being slowed up. The closer the circumference approaches the critical Schwarzschild value, the slower the implosion of the star. Finally, the collapse of the star freezes up—the implosion appears to have stopped.

NEWTON: That's fantastic! Gravitation becomes so strong that the time warp is not just enormous—it approaches infinity. In the limiting case, the flow of time grinds to a halt—just like a stream when the water freezes up all of a sudden. The star erects a wall of silence around itself—a wall of frozen time.

HALLER: For the outside world, the star vanishes, just like the Sun does for us at sunset—slowly at first, then ever more rapidly. The redshift of its light increases more and more. Finally normal starlight becomes some meager dull radio-frequency radiation.

In the limiting case, as the horizon has been reached—strictly speaking, at time infinity—no light is being emitted anymore. The light is being trapped by the strong gravitation of the star, which does not let it escape. That is why we call this a "horizon"—a definition that is clearly quite remote from our earthly horizon. The only way the star acts on the outside is now by way of its gravity; through its effect, the star's position can be obtained. It is only the external spacetime metric that can tell us that something is in existence inside. The star has turned black. That is the reason why a system such as this one came to be called a black hole—an expression introduced by John Wheeler; it is appropriate if a bit frivolous for this kind of a monster of spacetime, and was adopted by the physics community right away.

EINSTEIN: That reminds me of an old story which will interest Newton. All the time, you were of the opinion that light is composed of the smallest particles that move at a rate of 300,000 km per second, or 186,000 miles per second. This picture, from the vantage point of modern research, is not exactly correct, but it is also not wrong. Your idea was certainly popular in the seventeenth and eighteenth centuries.

Toward the end of the eighteenth century, the English naturalist John Michell had the following simple idea: assume we shoot small particles into space from the surface of a star. If the velocity of these particles is relatively small at the start, they will fly upward in the beginning, but then turn around soon because of gravity, and drop back onto the star. But if the initial velocity is sufficiently high— higher than some given escape velocity—it is also possible that the particle escapes into space. On Earth we know that this escape velocity is about 11 km per second.

NEWTON: It is a trivial task to calculate the escape velocity of every star or planet. Give me a minute. I arrived at 620 km per second for the Sun. The escape velocity depends only on two things— on the mass of the star and on its diameter. The larger the mass, the larger the necessary escape velocity. The smaller the radius for a given mass, the higher the escape velocity because gravity at the surface is stronger for smaller radii.

HALLER: That's where Michell started his reasoning. He left the mass of the star constant and reduced the radius until the escape velocity equaled the speed of light. This means he examined the case where the star's gravitation at its surface is so strong that even light particles—photons—can no longer escape. They are turned around by gravity and drop back onto the star's surface.

NEWTON: Just a minute, I'll get that right away. In the case of the Sun, the necessary radius amounts to 3 km. Eureka! The Schwarzschild radius is also 3 kilometers.

EINSTEIN: And that is no coincidence, Sir Isaac. If you look at your equations in detail, you'll notice that the mathematical expressions for Michell's critical radius and for the Schwarzschild radius are identical. The result depends on your gravitational constant and on the mass. On the basis of this simple reasoning, Michell arrived at the remarkable idea that there may be stars in the universe that are invisible because of their strong gravitational pull—they are simply dark. He reported this possibility in a lecture at the Royal Society in London and subsequently published his idea.

The French philosopher and naturalist Pierre Simon Laplace published a similar idea more than a decade later, but did not refer to Michell. We assume that he knew nothing of Michell's reasoning. At any rate, Laplace thought this was an important issue: he discussed his hypothesis on dark stars in his famous work "Le Systeme du Monde" (The World System), published in 1796. In the third edition of his book, however, this argument is no longer mentioned. Obviously, Laplace had meanwhile convinced himself that this argument did not make sense: your particle theory of light had been discredited in the meantime, and the new wave theory of light was slowly gaining acceptance. It took more than a century until the old formula of Michell and Laplace emerged again in a new guise, that of the general theory of relativity. It appeared custom-made by Schwarzschild, whose name—meaning "black shield"—appears well suited for the problem. A black hole is, after all, black. And it has a shield—its horizon.

NEWTON: It is good to see that my old theory is still good for something. But now enough of this historical talk, and back to the real world. What is the fate of our astronaut?

HALLER: He retains the possibility to escape before he reaches the horizon. That is the critical point for his return, assuming he still has enough fuel. If the astronaut returns to his colleague in the space station, he'll notice that time has passed more rapidly in the station than in his spaceship. By how much?—that depends on how closely he dared approach the horizon. There is no basic limit for the time warp.

NEWTON: The journey of the astronaut to the vicinity of the horizon might have lasted only a few days, but for the crew in the space station it took much more, say, fifty years. Is that true?

HALLER: Exactly. This would not really work for the black hole with ten solar masses that we talked about; but for a much more massive black hole, that would present no problem. Black holes are ideal machines for aging without growing older.

EINSTEIN: That sounds like an attractive possibility for middle-aged ladies; but let me stress, for real life there is nothing to be gained. The time passed close to the horizon merely helps the astronaut to pass gently into the future. There is no advantage to him personally—he lives in his time; everything that happens to him follows his proper flow of time. The black hole is not a fountain of youth. I prefer a walk here in the park to an excursion near a black hole.

HALLER: Let's leave that decision to the astronaut himself. Maybe he is very curious about finding out what happens at the horizon. If he does not flee from it in the last moment, he'll inevitably be drawn into the region of the horizon itself—and that means, he will be lost to the external world. There is no return—the black hole swallowed him.

EINSTEIN: There is something here that I don't understand. We just spoke about the fact that the collapse comes to a stop—and that would mean the strong time warp sees to it that the horizon will not really be reached. But in the system of the spaceship, time flows normally as always. The astronaut would not notice a thing as he approaches the horizon.

HALLER: That's not quite correct because, in the vicinity of the horizon, the effects of spacetime curvature grow very strongly. This means they will be noticeable inside the spaceship due to strong gravitational tidal forces. The astronaut, for example, will feel the the pull on his feet is much stronger than that on his head. How strong these effects are depends on the mass of the star in question. In our example, the effects are so strong that the astronaut could not survive in the vicinity of the horizon. He would be literally torn apart by gravitation.

EINSTEIN: Well, then we might have to do without the astronaut. We can replace the spaceship with a small space probe made of high-grade steel, which could resist such tidal forces.

HALLER: You are right. Let's take such a probe and imagine what happens as it approaches the critical point at the horizon. Now we get to the most interesting aspect of the black hole: Measured by the flow of time of the spaceship, the probe has an infinite lifetime. It survives when the astronauts in the spaceship have passed away due to old age. Let's say the probe is now located just above the surface of the collapsing star, which is getting close to the horizon. This is where the relativity of time really hits. Inside the probe, there is obviously no noticeable effect of a slowing of the flow of time, which the crew of the space station might observe. The probe dashes in the direction of the star's origin, together with all stellar matter. Its velocity, as measured by its own flow of time—its proper time—is increasing all the time, until it reaches the speed of light at the horizon. The probe zooms through the critical point without any special effect.

NEWTON: Just a moment, please! How can that be? We had seen that something does happen at the horizon when we take the Schwarzschild solution of Einstein's equations. The flow of time relative to the observer at rest simply stops. The Schwarzschild solution makes no sense inside the horizon.

HALLER: Let's first address the question whether something drastic really has to happen with the probe at the horizon. The answer is:

no. The solution of the Einstein equation given by Schwarzschild applies in a given reference system, where an observer at consider-able distance is at rest. It turns out that this reference system is simply not capable of describing what happens inside the horizon. That is nothing unusual in the general theory of relativity. Quite often, we are able to describe only some given regions of spacetime in terms of the chosen system of reference or of coordinates. If we want to examine the missing part of spacetime, we have to come up with another coordinate system. This changes only the description, not the actual physical truth—and that is what matters. It is not the metric tensor that has physical importance—it is the tensor of curvature.

If you start from the Schwarzschild **ansatz** for the metric tensor and determine the tensor of curvature from it, you'll notice that nothing dramatic happens at the horizon. The transition from the reference system where the space station is at rest to that of the moving probe is just that—a change of reference systems. And in this new system, we notice that the curious happenings we found in the Schwarzschild solution vanish as the circumference approaches the Schwarzschild circumference.

The relevant calculations were done, incidentally, by Oppenheimer and his collaborator Hartland Snyder a short time before the beginning of World War II. This was Oppenheimer's last scientific project before starting on the Manhattan Project.

EINSTEIN: Robert Oppenheimer then found that the probe passes through the critical point, still following the surface of a collapsing star. Seen from the vantage point of the observer in the space station, this happens only after an infinitely long time has passed—you might say, in an infinitely remote future. For the probe itself, on the other hand, the matter takes just a fraction of a second. But what happens to our space probe after it passes the horizon?

HALLER: Immediately after passing through the critical point, the probe has lost all possibilities of contacting the space station. If it emits radio signals, these signals will be deflected by gravitation such that they remain inside the horizon. Light rays from the outside, on the other hand, have no trouble entering. The probe acts a bit as a valve—matter and light can enter, but they cannot exit. Whatever crosses the horizon is lost.

EINSTEIN: We should place a large sign at the horizon with the inscription that Dante put at the entrance to hell, in his *Divine Comedy*: "Bid hope farewell, all ye who enter here."

HALLER: That would be well justified. Although the horizon is still harmless, we'll see that its job is only to disguise the "hellfire" in the center of the sphere. Our probe will soon be confronted with it as it approaches the center of the black hole. In our example, it takes only a few time units in the Schwarzschild frame before the probe hits the very center point, just behind the collapsing stellar matter.

NEWTON: Does that mean that the entire stellar matter has dropped into the center, into just one point? If that is so, I think we really have a problem.

HALLER: You can say that again! What I described is the result of calculations. All stellar matter collapses into one point. When Oppenheimer realized this, he was unable to come to terms with his own results. He refused to think of the consequences—he actually ran away from them and left the matter to its own devices.

EINSTEIN: I don't want to have it appear as though we did the same—just run away from them. But as much as I would like to see the solution of this problem, it doesn't make sense. If we still want to catch lunch at the Athenaeum, we have to hurry. Well then, gentlemen, let's get going! I certainly need something to eat. After lunch, we can pay a visit to the bottom of the black hole. I insist, rather egotistically, that we do this only in theory.

CHAPTER 16

In the Atrium of Hell

The greatest of miracles is this: there are no miracles.
—Albert Einstein[1]

For lunch, Haller had reserved a table in the corner of the Athenaeum dining hall. Still, there were a few faculty members that were ogling the small group. But our threesome was not to be bothered—in particular, Newton had no intention of waiting until the afternoon to learn more about the center of the black hole. "One thing is not to be contested" he started; "the collapse of stellar matter into a single point, which follows from the calculations based on spherically symmetric collapse, must be absurd."

HALLER: There I'll agree with you. We have to deal seriously with the question whether the fundamental assumption for this calculation—that of strict spherical symmetry, or even Einstein's equations (sorry about that, Mr. Einstein!), can be justified in a situation as extreme as this one. Just these questions were asked by physicists, but only long after Oppenheimer's work on the subject. It was not before the 1960s and 1970s that physicists worked again on the problem of the gravitational collapse of heavy stars into black holes. There is no conclusive clarification to date.

Let me tell you in a few words what the state of that investigation is today. Let's begin with one thing: whatever is gobbled up by a black hole will be totally destroyed as it plunges into the center—not only its macroscopic structure but also the atoms, the particles that make it up.

If you take the result of the calculations seriously, then the entire stellar matter, including the material that made up our probe, will become concentrated in the center of the black hole—in just one

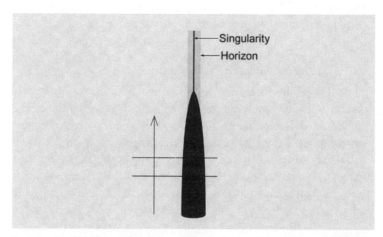

FIGURE 16.1 This figure shows the collapse of a massive star into a black hole. In the process, stellar matter first pierces the horizon, then the center, forming a singularity in spacetime.

point with infinite matter density. The calculations result in something mathematicians call a *singularity*. In the case at hand, the term applies to matter density and spacetime curvature. Since curvature describes gravitation, we might say that as we approach the singularity, the gravitational interaction will also grow to infinity. Neither space nor time retain the form we are accustomed to.

EINSTEIN: It is not the horizon that is the problem of the black hole, although it may look that way at first! No, it is its center that gives us trouble. We might say that it is at the origin, in the first Ring of Hell, that we find the devil.

HALLER: You can say that again. If you are looking for a physical definition of hell, this is it, right in the center of a black hole.

NEWTON: It is interesting that Nature obviously is ashamed to show this fiasco openly. For that reason, Nature surrounds the singularity with a veil—the horizon. Everybody who goes to hell first has to go past the horizon; as he does so, he loses all contact with the outside world. The return ticket, you might say, is taken away. As soon as he actually sees the singularity, he is lost; there is no way back.

HALLER: The question is whether the singular point in the center of the black hole is only a property of the mathematical solution of the problem, or whether the singularity is actually realized by Nature. It is, after all, possible to imagine that spherical symmetry

FIGURE 16.2 The British astrophysicist Stephen Hawking. (*Photo:* Cambridge University)

will be violated as all that stellar matter drops down into the center. It might be that part of the matter could drop down in a slightly different direction. As all that stellar matter is precipitated toward the center, all hell breaks loose anyway. Chaos prevails, so that is not an unreal scenario. And that creates a new problem—we can no longer show that there is a strict singularity. During the chaotic process of the fall toward the center, the singularity might be dissolved as such.

But we are not sure that this could actually happen. It could be that the dynamics of gravitation, such as Einstein's equations describe it, do not permit this dissolution of the singularity: it might, instead, stabilize the collapse toward the center; it might dampen small deviations from spherical symmetry, ultimately straightening them out.

Singularities in spacetime are nothing unusual in the framework of Einstein's theory. At the end of the 1960s, two British mathematical physicists, **Stephen Hawking** and **Roger Penrose**, showed that the dynamics of spacetime as described by Einstein's equations will always lead to singularities, provided there is matter present. In the absence of matter, this, of course, doesn't apply. That is shown by the example of planar spacetime, where there is neither curvature nor matter. As soon as there is matter, there will always be event points in spacetime, and that leads to singularities. The singularity of the black hole is just one example. We will later see that the origin of the universe in a Big Bang is another example.

EINSTEIN: This would seem to argue in favor of the idea that, when there is a violation of spherical symmetry, the singularity cannot be avoided. I cannot but admit that I myself am astonished by all these consequences of my own equations. It looks like I awakened ghosts I can no longer get rid off.

But let me warn you—do not take my equations all too seriously. In a situation as extreme as the collapse into a black hole, it might be possible that the equations turn out to be nothing but very good approximations of a more comprehensive, unified description of gravitation and matter.

HALLER: May your word rest in God's ear, Einstein! I shall remember your remark, and I suspect that soon enough I'll have to remind you of it. Most physicists who deal with gravitation these days think that the problem of singularities in black holes occurs because we deal with Einstein's equations as though they were equations of classical physics. We neglect quantum effects. But quantum physics comes in as soon as you look at the dynamics of matter at very small distances—say, on the atomic or particle level. That means we cannot investigate the effects of gravitation without sooner or later discussing quantum mechanical effects.

In Einstein's equations, we have the tensor of spacetime curvature on the left side, a purely geometrical quantity; on the right side, there is the matter tensor. It is indisputable that matter has quantum properties. But they don't even show up in Einstein's equations. The matter tensor as it shows up in the equations is not the *ultimate* matter tensor, which also has quantum properties: rather, as shown in the equations, it has been subject to some averaging.

Since matter consists of particles and, therefore, has a somewhat grainy structure, we have to "smear" small regions of spacetime to make this granularity invisible. Take, for illustration, a TV picture: it consists of many, many tiny points—but at a certain distance these are no longer distinguishable. As we look at the screen, we don't see the granular structure. The eye "averages out" several points and registers only a continuos image.

In the same way, the matter tensor enters into Einstein's equations in a continuous form that averages over small discontinuities. Spacetime reacts to matter just as our eye reacts to the granular television image—it averages out over small areas. And that is fully good enough for the investigation of gravitation where matter accumulates on a large scale, as in the case of a star.

But problems pop up when we study phenomena where small

dimensions in spacetime become important. And that is exactly what happens when a star collapses into a black hole. That's when the singularity comes into the game.

NEWTON: I will certainly agree with that. And now I think the time has come that we have to confront the quantum properties of gravitation.

HALLER: You say that almost in passing, as though we could take care of it over lunch. In reality, an attempt to do so will meet with almost insurmountable difficulties. And these are not just problems of a mathematical or formal nature—they require in fact a new version of the concepts of space and time.

Gravitation is a manifestation of the curvature of spacetime. The latter, in turn, is determined by matter via Einstein's equations. Since matter has quantum properties, we are forced to grant such properties to space and time also if we want to maintain Einstein's equations on the level of quantum physics, too. But that has not been possible to date. Nobody knows how to treat space and time quantum mechanically: the dynamics of quantum processes, for example atomic reactions, after all, happens in space and time. Now, if we ascribe quantum properties to space and time, we would have to destroy the very basis of what we wanted to build on. We find ourselves in the situation of a developer who just finished his new house, only to discover that he built it on marshland.

But one thing is fairly certain: our customary concepts cannot be maintained when spatial or temporal dimensions become very small. The manifestations of quantum physics, such as the structure of atoms, are determined by a constant that Max Planck introduced to physics; it is usually denoted by h. It has to be determined experimentally. Its precise value does not matter here. What does matter is the fact that it pins down the sizes of atoms, such as the diameter of the hydrogen atom, which amounts to about 10^{-8} cm. The strength of gravitation is determined by Newton's constant G. This constant describes, as we know, the interaction between matter and spacetime in Einstein's equations. Max Planck noticed at the beginning of the twentieth century that it is possible to fix small intervals of both length and time with the aid of his own new constant of nature, of the Newtonian gravitational constant, and of the speed of light. They are often referred to as

Planck's elementary unit of length (Planck Length) and of time (Planck Time). Let's look at them more closely:

Planck's elementary unit of length	$1{,}616 \times 10^{-33}$ cm
Planck's elementary unit of time	$5{,}391 \times 10^{-44}$ seconds

The elementary length and elementary time are connected by way of the speed of light. The elementary time is the time it takes for light to travel a distance that equals the elementary length.

NEWTON: I feel honored. My gravitational constant actually leads to the appearance of a specific length in quantum physics. That means gravity has something like its own scale. However, it is so tiny that it is below any imaginable value—it is 25 orders of magnitude smaller than the diameter of a hydrogen atom. What then is the physical significance of these elementary units? Are they the quantities that determine the quantum properties of space and time?

HALLER: I wish I could give you a precise answer. At any rate, the Planck units are of considerable interest. They are the only ones that can be derived by means of fundamental constants. All other units of length and of time that show up in physics, are tied to macroscopic—and therefore coincidental—quantities; or they are tied to certain properties of particles. What roles the Planck units really play, we still don't know. We surmise that at small distances in space and time, at distances of the order of the elementary units, space and time will be subject to well-known quantum physical uncertainties. But this is only an assumption, nobody knows the details.

EINSTEIN: I believe this will not change in the future. As far as I'm concerned, I don't think there is any reason to drag space and time into the quagmire of quantum physics. I admit, of course, that quantum physics is utterly important in atomic physics. I was among those present at its baptism. Still, I warn you against subjecting my theory of gravitation to the hypotheses of quantum physics. A forced union like this will not lead any place.

As you mentioned, the consequences would be that space and time have quantized properties—a thought that makes me shudder. Space and time are the foundations of all natural science—they are the scaffold that props it all up. And now the quantum physicists come along and say that this scaffolding is built in a mushy quantum quagmire. Leave me out of this one! I will admit that my equations become questionable as soon as there are phenomena as singular as the gravitational collapse into a black hole. What we need is a comprehensive theory that unifies gravitational theory and quantum physics. I have no objections to that. But I will resist today's fashion that wants to subject my equations to the laws of quantum physics. That would be a colonialization of the general theory of relativity, and certainly not a unification. A quantum version of my theory— should one exist—would be a horror to me. I think Nature agrees with me on this one.

HALLER: There is nothing I would like less than entering into an argument about the part played by quantum physics in gravitational theory, and that at lunchtime to boot. Let me, instead, get back to our basic problem. We surmise—and most physicists today agree with me on this one—that the problems we detected in conjunction with the gravitational collapse into a black hole will come into play once the extent of the collapsing matter is no longer large when compared with Planck's elementary length. We might imagine that the collapse will then stop. That would avoid the singularity. There would be no infinities; they are being ironed out, if you wish, by quantum physics.

NEWTON: I will accept that there is no singularity, in a strictly mathematical sense. But let's note that when we smear all that stellar matter on to a truly minuscule sphere—a sphere that has Planck's elementary length as its radius—there will be a matter density and an energy density of hardly imaginable proportions. So what is this matter that we are talking about? It could not be either nucleons, protons, or neutrons, nor quarks. So what kind of soup is the black hole cooking at its center?

HALLER: It looks as though, today, that I am running away from questions rather than answering them; but I have to repeat that, for this question, neither I nor anybody else can as yet answer it. We know that, under normal conditions, matter consists of the quarks that make up atomic nuclei, and of electrons. That goes for the collapsing star too, as it starts its catastrophic sucking motion. It is questionable that under the extreme conditions that prevail in the center of the black hole, the normal concept of an elementary particle, such as we use it in particle physics, is applicable at all.

EINSTEIN: You see, not only the concepts of space and time need to be revised, but also those of particle physics and quantum physics. I could get used to that. I mentioned before the need to unify the theory of gravitation and of quantum physics. Should that work, there will be no way to separate matter and gravitational phenomena in this unified construct. Space, time, and matter will manifest themselves as different aspects of one and the same basic phenomenon. I can imagine that at the center of the black hole, this unity will finally prevail. But how that would happen I have no idea.

NEWTON: You mean that in the center of the black hole there will be some kind of "ur"-state of space, time, and matter?

HALLER: Einstein could be right. We'll get to cosmology in the next few days. That is where we will find out that there was another such "ur"-state of matter and spacetime right after the birth of our universe—but it persisted for the briefest of times only.

But let me suggest that we have discussed the nature of the singularity in the black hole for all it's worth. We have gotten as far as today's knowledge has advanced. Nobody knows how long it will take before we know what actually happens at the very center of a black hole.

EINSTEIN: To the outside, a black hole manifests itself by only one characteristic: its mass. Or is there anything else?

HALLER: A black hole, such as we have contemplated it, and as the Schwarzschild solution to Einstein's equations describes it, has no further property. It is fully characterized by what we know about its mass. Note that it may also have an electric charge should the matter that dropped into it during stellar collapse have been electrically charged, but we shall disregard that. Two black holes of equal mass are totally identical. A number of astrophysicists used a drastic expression for that. They said a black hole has "no hair."

EINSTEIN: That might be correct, but I have a problem with it. As we know, there is not only matter but also antimatter. Particle collisions are capable of producing an antiparticle for every particle there is. True, our cosmos doesn't appear to contain large amounts of antimatter, but that is not the point here. Let's assume we have a star consisting of antimatter—antiprotons, antineutrons, and so on. Now let it collapse, and nothing remains but a black hole. Your "no hair" theorem would say that there is no difference between a hole that formed by the collapse of matter and another one that is due to the collapse of antimatter. Is it true then that we cannot trace a black hole back to its origin?

HALLER: I wouldn't know how. But I'll admit that this is another unanswered question. If there is truly no difference between the two types of black holes that we discussed, then that would mean there is no difference between the "ur"-state in the center of the black hole, be it due to matter or antimatter collapse. But let's leave speculation aside and turn our minds to tangible results.

In the 1960s, further solutions of the Einstein equation were discovered; these solutions correspond to a rotating black hole. Such an object revolves about an axis. It doesn't have a spherical horizon. The horizon, instead, shows some flattening at the poles, just as the Earth does. The larger the velocity of rotation, the more obvious the flattening. That means that every black hole can be characterized by its mass and its rotational velocity. There is no other mark it carries.

If a piece of matter drops into a black hole, its mass will increase, and so will, ultimately, the velocity of rotation. But there is no way that the black hole can give information about the matter it gobbled up. It makes no difference whether it was a gas, a steel beam, or an astronaut that consisted mostly of hydrocarbons—a black hole has no trouble swallowing up all that matter—even if it were antimatter.

EINSTEIN: My stomach reacts differently. After this big lunch I need a good siesta.

Einstein left the dining room. Newton and Haller waited a while before taking a walk along California Avenue to where it crosses Lake Street. And there they settled down at a table in the latest coffee-craze establishment.

CHAPTER 17

The Monster of Spacetime

The untiring efforts of the researcher are motivated by an overpowering secret urge: what he wants to understand is nothing less than our existence, our reality.

—Albert Einstein[1]

At three o'clock in the afternoon, the meeting continued in the small library of the Athenaeum. When Newton and Haller arrived, Einstein had already penetrated the secrets of the coffee machine that was in the room and was now ready to start pouring. Newton opened this round of conversation: "Up to this point, our discussions on black holes have been purely academic. But one thing is certain: black holes are legitimate solutions of Einstein's equations. First, we have to ask: are there any black holes in our universe."

EINSTEIN: Just a minute, Sir Isaac. Before we ask Haller for an update on this issue, let's think about how one would go about looking for a black hole. And looking from Earth, you understand, not from some hypothetical spaceship that starts out on a kamikaze flight toward the horizon of a black hole.

NEWTON: A black hole influences its surroundings mainly through gravitation. Unfortunately, that is not a particularly conspicuous criterion for an observational astronomer. To observe a black hole in the dark universe—that sounds like a difficult task to me, as hard as photographing a black cat as it goes mouse-hunting at night.

EINSTEIN: I'm not quite that pessimistic. There are many two-star systems in the universe. When one of these stars implodes to become a black hole, we will be left with one star and one black hole that orbit each other. We would have to examine two-star systems where one partner is not visible except by the influence it exercises on the orbit of its visible partner.

HALLER: Originally, this was expected to be a promising method for finding black holes. Particularly in the Soviet Union, astrophysicists specialized in this method; but this method turned out not to be very practical, because it is not easy to really identify the black hole—should there be one in the first place. The situation would be much helped if the black hole could be pinned down not only by means of its gravitation but also by at least one further effect.

NEWTON: But the black hole is black—it can't be seen in a telescope; so I guess we can forget that one.

HALLER: A black hole need not be all that black. Let's assume we have a system consisting of a black hole and a normal star that orbit each other. Some stars constantly emit large amounts of gas into the universe. This is true for our Sun, but the gas the Sun emits in the form of solar winds is limited in amount. The black hole might move across stellar winds of its neighboring star. Part of this gas would be sucked into the black hole. But most of the gas molecules, while being attracted by the strong gravity of a black hole, would then bypass the horizon and, in the process, this would lead to strong density variations and vortices in the gas. The gas could also form shock waves that heat it up strongly, and produce beams of X-rays.

EINSTEIN: Excellent—that gives us enough of a profile to put our black hole on the Ten Most Wanted list. So, we have to look for two-star systems consisting of a normal star and an invisible companion that emits X-rays. Technically, we have a bit of a problem here: X-rays, unfortunately, are absorbed by the Earth's atmosphere.

HALLER: That is a good thing, I should think. Otherwise, we would be exposed to constant bombardment by X-rays. But as of the 1960s, we have been searching the skies for X-ray sources by means of rockets and satellites. In 1978 the United States launched a large X-ray telescope that was given your name—it was christened the Einstein telescope. Today, we investigate the X-ray skies mainly with the help of a satellite called Rosat.

Right now, we have neither the time nor the means to discuss the details of X-ray astronomy. But this field has become an important and exciting part of modern astronomy. Let me just say that what matters here is the fact that hints were found for the actual existence of black holes. The most interesting candidate we have is the X-ray source Cygnus X-1, an object the mass of which we have not found out precisely, but it has to be somewhere between 3 to 16 solar masses—well in the range that theory has determined to apply for black holes. Its companion is a normal star the mass of which we roughly know to have between 20 to 34 solar masses.

Notwithstanding the large uncertainties that remain for these masses, the details of what we see in the Cygnus X-1 system make it very likely that we are dealing here with a black hole. Experts peg the likelihood that this is so at 90 percent. We don't have a proof that will make it 100 percent, and we may never get that. There is no absolute certainty here, just as in many other fields of the natural sciences. In the meantime, we have knowledge of many X-ray sources similar to Cygnus X-1, so that we have put together an impressive "black list."

EINSTEIN: If you consider that, at least in theory, many stars have enough mass so that they might wind up as black holes at the end of their stellar lives, it is clear that the number of black holes in our galaxy could be considerable.

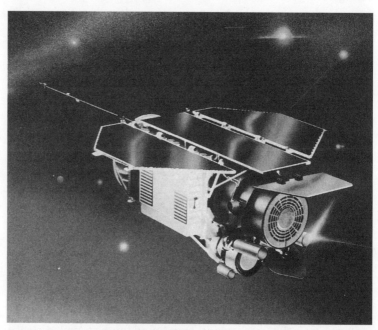

FIGURE 17.1 The X-ray satellite Rosat, instrumented to scour the sky for X-ray sources. (*Photo:* Max Planck Institute for Extraterrestrial Physics, Garching, Germany)

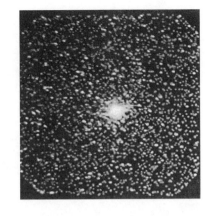

FIGURE 17.2 Mapping of X-ray sources in the sky, prominently showing the X-ray source Cygnus X-1, which presumably contains a black hole. (*Photo:* NASA)

HALLER: Some astronomers believe that the number of black holes could be comparable to that of the optically visible stars, and maybe even bigger. Black holes—in contrast to massive stars that can look forward to long lives on a cosmic time scale—are dead objects. They have no active careers to look forward to; you might call them stars emeriti. But they will not really die. In the long history of the universe, there were certainly many stars that had at least a chance to collapse into a black hole.

NEWTON: There might be one hundred billion black holes in our galaxy alone—isn't that absurd? A large part of the mass our galaxy will then consist of black holes.

EINSTEIN: I remember what Fritz Zwicky said a long time ago here at Caltech. He claimed that in the galactic clusters that he was studying—for instance, the Coma cluster—the visible matter was not sufficient to explain the dynamics of galactic motion. He found that a large part of the matter was missing, that it was, so to speak, invisible. Maybe that's the hint we need for black holes—always supposing that Zwicky's observations were correct.

HALLER: I know well that most colleagues at the time did not take Zwicky's remarks very seriously. But later on it turned out that he was correct. Both in individual galaxies and in galactic clusters, a large part of matter is invisible. Today, there is talk of missing matter. Black holes could be part of this missing matter. But it is unlikely that the missing matter consists exclusively of black holes. We have good reason—which I will not discuss right now—to infer that only a small part of the missing matter could be accounted for by black

FIGURE 17.3 The center of the Coma cluster of galaxies (in the region of the constellation Coma Berenices).

holes. But that does not exclude that the number of black holes in our galaxy alone might be of the order of ten to one hundred billion.

EINSTEIN: If we look at the catalogue of possible black holes, I'm sure we should also find extremely massive ones. What is the mass of the most massive candidate we have so far?

HALLER: Your question puts me on the spot. I have to admit that I had not yet mentioned what is probably the most interesting aspect of the black hole. But, since you asked directly, I would like to give you a direct answer: it is likely that somewhere in the universe there are black holes with a mass a billion times that of the Sun.

NEWTON: What? Billions of solar masses? Up to this point, we were always talking of masses of the order of a few solar masses. If what you say is correct, we must be talking about objects that are qualitatively different, that have truly galactic dimensions—they must be monsters in spacetime.

HALLER: That is precisely what they would be—nuclei of galaxies. Let me quickly describe how we got to the discovery of these objects. It has been known since the 1930s that the center of our galaxy is a source of intense electromagnetic radiation. In the 1950s, cosmic radiation sources were discovered that appeared to be pointlike, like individual stars. The name of these objects was chosen accordingly: we call them quasars, a shortened version for quasi-stellar radio source. Later it was discovered that some quasars emit not just radio waves but also visible light, and that made them look like normal stars. But when we made it our business to examine the light of these stars more closely—and that was largely done here at Caltech—there was a surprise: the distance of these objects from the Earth is of truly cosmic dimensions—in fact, billions of light years.

NEWTON: If these quasars are some sort of stars, they cannot be that far away. How else would they become observable on Earth? We can see galaxies at these distances, but not individual stars!

HALLER: Just that fact was a surprise. These objects looked like normal stars of our galaxy; but then they turned out to be objects much farther removed; and the energy they radiate is truly gigantic. There are many quasars that emit more than a hundred times as much energy as our entire galaxy. And all this energy comes from a limited region in space. That region cannot be much larger than our solar system.

There is just one mechanism that is capable of producing such gigantic energies both in the region of visible light and in that of radio waves, and that is the collapse of gigantic amounts of stellar matter into a black hole. In order to produce the energies we observe, we need black holes with masses of the order of a billion solar masses. The Schwazschild radius of such a black hole is about one billion times larger than the Schwarzschild radius of the Sun. It will amount to some three billion kilometers, or 10,000 light seconds. The Schwarzschild radius of such a massive black hole is about one light hour. That is comparable with the radius of our solar system. And this explains why quasars appear pointlike in our telescopes.

EINSTEIN: I have trouble understanding just how such monstrous black holes can come about.

HALLER: The dynamics of solar motion in a galaxy—in particular, the effects of friction and gas clouds emanating from that—sees to it that more and more solar matter will assemble in the center of the galaxy, including both stars and huge clouds of gas. That may well cause the gravitational collapse of the entire central region, and thereby the emergence of a gigantic black hole. It could also be that several smaller black holes form individually, but later on merge together. Be that as it may, in the end there will be just one extremely massive black hole.

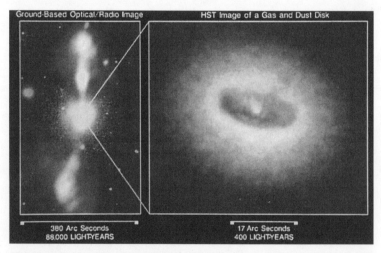

FIGURE 17.4 A quasar, the energy of which is presumably supplied by a very massive black hole at its center. (*Photo:* NASA)

EINSTEIN: If that is so, I would expect a black hole to be present in the central region of all large galaxies; and this would be true for our own Milky Way as well.

HALLER: That is just what we expected. In quasars, the central black hole is surrounded by a lot of matter. That is the reason why there is a gigantic release of light energy. Astronomers are actually being blinded by the superluminous center and so cannot even see the actual galaxy.

Other galaxies have much less matter near the center. This is true particularly for older galaxies where the central black hole has, so to speak, gobbled up everything that was in the central region. And that is why there is no longer a strong source of light emission in that region.

Chances that a very massive black hole will form in the center of a galaxy are greater the more massive the galaxy is. An interesting object in this respect is the galaxy we call M87; it is the most massive galaxy in the universe that we can see. Gigantic and almost spherical, it is located close to the center of the Virgo cluster; we have good reason to believe its mass exceeds that of the Milky Way more than fiftyfold. Its distance from the Earth is about 40 million light years. Its most conspicuous feature is that there is almost no matter in the form of gas clouds close to its center. But the center does emit intensive radio waves and X-rays. All indications, then, are that its center contains a very massive black hole.

EINSTEIN: Our own galaxy has no very conspicuous features and is none too massive when compared to others. Are there any indications that it contains a black hole in the center?

FIGURE 17.5 The galaxy M87 in the center of the Virgo cluster of galaxies; it most probably contains a black hole at its center. This is the most massive galaxy known in the observable universe. The galaxy is surrounded by many small star clusters, some of which are visible in this image. (*Photo:* Hale Observatory, Pasadena)

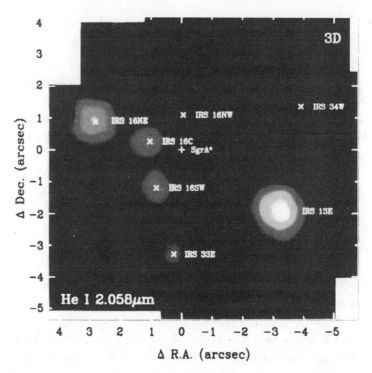

FIGURE 17.6 Central region of the Milky Way. The radio-source star SgrA* on the left side above the center is a candidate for a black hole with a mass in the region of between 2 to 6 million solar masses. (*Photo:* Max Planck Institute of Extraterrestrial Physics, Garching)

HALLER: More recently, the attention of astronomers has been directed toward the galactic center. There are indirect indications that the Milky Way does have a black hole there; but its mass is likely to be small in comparison to others—only about three million solar masses. That would make the Schwarzschild radius of this hole about ten million kilometers, or one half light minute—about one twentieth of the radius of the Earth's orbit in our solar system.

EINSTEIN: Provided it is correct that many galaxies have voracious black holes at their center, what does that indicate about their future? Does that not imply that, in the course of time, more and more matter will vanish in there until, in the end, what remains of the whole galaxy is just a black hole? This fate would then be shared by the atoms you and I are made of—not a particularly appetizing prospect for the future!

HALLER: That does not have to be. As luck would have it, we are living in a stellar system pretty far removed from the center—by about 30,000 light years. A black hole that gobbles up almost our entire galaxy would have to have a Schwarzschild radius of just one light year; and that is a good deal less than the radius of the orbit of our Sun.

EINSTEIN: What is the time scale for the cosmic gluttony that we ascribe to the black hole?

HALLER: We estimate it will take about 10^{17} to 10^{20} years; that is at least 7 orders of magnitude longer than what we estimate the age of our galaxy to be.

EINSTEIN: That does sound comforting. On the other hand, who knows what else might happen to the orbit of our Sun, which we expect to cool down in as little as five billion years, a short span of time in comparison with what we said about the black hole. Casual "close encounters" with other stars could change its orbit considerably.

HALLER: You are certainly right. It is unlikely that the solar orbit will not change. That is the reason why we cannot predict what the future of our atoms will be. Their chance of ending up in black holes at the galactic center is not small. However, one comfort is the time scale: 10^{17} years is a long time.

EINSTEIN: Before we leave off for dinner, there is another question that has been bothering me ever since we started talking about black holes: according to my theory of gravitation, a black hole is a time-independent solution of the equations, with a singularity at its center. But we have already noticed that the singularity is probably an artifact of my equation. It is most probably going to be smeared by quantum effects; that would eliminate the infinities. But there is another infinity that bothers me. According to my equations, a black hole is absolutely permanent. Once it has formed, it remains forever. It is absolutely stabile. But I have doubts whether this can really be, once quantum effects come into the game.

Here is an example from nuclear physics. You know that some atoms are unstable, but with lifetimes that may amount to thousands of years. That means that a nucleus decays, emitting several other particles in the process. If you examine it more closely, you will notice that the decay cannot actually happen according to the forces at hand, and the laws of classical physics. But it does happen;

quantum physics makes it possible, even if it takes a long time. Nature digs a tunnel to outwit classical physics. We speak of the "tunnel effect."

A black hole is black only to the outside; but as we said before—all hell breaks loose in its interior, given its enormous energy density. I can imagine that Nature finds a way to get rid of at least part of this energy in some ingenious way similar to, for instance, the tunnel effect.

HALLER: You mean that there could be a kind of tunnel effect for black holes, which makes it possible that black holes are not absolutely stable after all? This is exactly the direction I wanted the discussion to take before we finish talking about black holes. So, why don't we discuss this now, even at the danger that we'll have to start dinner a bit later than usual?

The Soviet physicist Yakov Zeldovich noticed in the early 1970s that quantum physics may influence the dynamics of black holes. He considered rotating black holes and noticed that the velocity of rotation will slow down as time proceeds. In the end, a nonrotating black hole will remain. In the process, it will emit energy in the form of radiation. Zeldovich found this out by studying, as an analogue, a rapidly rotating metallic sphere. According to the laws of classical physics, this sphere, if it is somewhere in outer space, will not be subject to any frictional effects, and that means it will keep rotating forever, since there is nothing that could brake the rotation. If you add quantum effects to the calculation, this cannot be. It turns out that the rotation of the sphere will stop in due course; its energy will be emitted as electromagnetic radiation. But the effect is very small, and we have no experimental proof to date. It is just a calculation in the framework of electrodynamics.

EINSTEIN: I can imagine where the effect originates. Although the metallic sphere rotates in empty space, we have already seen that this space is not really empty—it is replete with all kinds of virtual particles. That makes me assume that our sphere will encounter friction with this "gas" of virtual particles.

HALLER: You might put it that way. Among these virtual particles, there are also photons, particles of light. Some of these photons take advantage of the existing rotational energy to borrow some of that energy, in order to change their virtual existence and become real particles instead—emerging as electromagnetic waves. The result is that our sphere emits radiation while losing more and more rotational energy. The sphere slows down.

Zeldovich's analogue ran into opposition from all the black hole experts. Striving to prove him wrong, they found that Zeldovich had examined only part of a much more far-reaching effect. The first to discover that nonrotating black holes also emit radiation was Stephen Hawking in England. Hawking, by the way, holds Newton's chair of physics at Cambridge University today.

EINSTEIN: I had a premonition that there must be more. So there is a kind of tunnel effect for black holes.

HALLER: The analogy with the tunnel effect does not really help, although the radiation we discussed is due to a similar effect in quantum theory. Normally, virtual particles play no part in the phenomena of classical physics. But gravitation is very strong near the horizon of a black hole, and that has consequences for the ocean of virtual particles. Hawking established that a black hole is capable of pulling virtual particles from the vicinity of the horizon into the black hole proper. We have already seen that virtual particles can pop up in pairs from the vacuum, say, an electron-positron pair. One of these particles has negative energy, the other one has positive energy, and the sum is zero, just as it should be. When this process happens close to the horizon of a black hole, something odd can occur: the particle with negative energy can be pulled into the black hole, never to return. But its positive-energy partner may stay outside the horizon. It is now alone, and will be forced to continue as a real particle. The particle that was pulled into the hole gives up its negative energy in the process. The hole has no choice but to register this energy loss, sort of like a bank when it has accepted a bad check. That means the black hole loses a tiny amount of energy—just as much as the energy that the other particle which gets away as a real particle, will take along.

When Hawking examined this more closely, he found that the particles that are radiated off have a given temperature. That means a black hole acts a bit like a heater that gives off heat in the form of radiation. We can understand the origin of the radiation of a black hole in the following way: a classical black hole has an exactly defined horizon. A particle that is outside the horizon by as little as a fraction of a millimeter is still safe; but a particle right next to it that is inside the horizon is lost. The horizon is really a sharp dividing line. But quantum physics says that such a limit cannot be arbi-

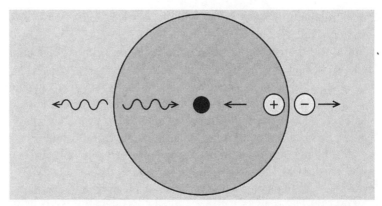

FIGURE 17.7 A black hole is capable of tearing virtual particle-antiparticle pairs apart so that one of the two vanishes beyond the horizon. As a result, there is an energy loss as particles vanish in the process.

trarily well defined. Just like all atomic processes, whatever happens close to the horizon is subject to the quantum-mechanical uncertainty. The location of the horizon cannot be fixed with arbitrary precision, it fluctuates over a small region. That means that a particle close to the horizon will not a priori know whether it will be sucked into the hole or not. We can only state a probability for that to happen. There is also a certain probability that a particle, say a photon, will get away from the black hole and take its energy along with it. In this way, the black hole radiates off energy.

EINSTEIN: So, a black hole is not all that black—it is warm. The radiation that it emits is like heat radiation. But what is its temperature? A nonrotating black hole is characterized by only one number, its mass. That means its temperature must depend on the mass, or on the Schwarzschild radius, which in turn is fixed by the mass. I assume, then, that the wavelength of this radiation is somehow tied to the Schwarzschild radius.

HALLER: Your assumption is pretty good. The wavelength of the radiation corresponds, in fact, to the order of magnitude of the Schwarzschild radius. A black hole that has ten times the mass of the Sun would emit electromagnetic radiation with a wavelength of about 30 kilometers.

EINSTEIN: This is not a small wavelength; it is much greater than that of radio waves. And that makes the temperature of the hole exceedingly low.

HALLER: Yes, it is smaller than a one-millionth part of a degree. And that also makes the radiation of energy exceedingly small— and the lifetime of the black hole correspondingly long. It will take 10^{69} years until the mass of our black hole has been radiated off. The lifetime of other black holes we can easily determine if we recall that it is proportional to the third power of the mass. A black hole with a mass of one billion solar masses would have a lifetime of 10^{93} years.

NEWTON: That puts us in competition with infinity. I would say that, for physics and astrophysics, these lifetimes mean little—for all practical intents and purposes, the black hole is stable.

EINSTEIN: I'm more concerned with the principle here. Suppose we were able to wait out such long periods of time: what finally will happen to the black holes when the mass has been used up?

HALLER: When the mass is significantly smaller, the temperature rises accordingly. This means that radiation becomes more effective and that the reduction of the mass accelerates. Finally, the Schwarz-schild radius of the black hole has a dimension of an atomic nucleus. But a hole like that is no longer black—it is red-hot. A large fraction of the radiation will now be emitted as X-rays. The mass decreases even more rapidly. Finally, there is an explosion that frees amounts of energy on the order of a million hydrogen bombs.

EINSTEIN: After this explosion, the mass of the hole has changed into energy. Does that mean that nothing is left of the hole?

HALLER: Probably, but we are not sure. The problem is that the end phase of the explosion cannot be calculated—we are back to the quantum aspects of gravitation. As soon as the Schwarzschild radius has the dimensions of Planck's elementary length, all bets are off.

NEWTON: Please note that we are now concerned with things that might happen, if at all, in an exceedingly distant future.

HALLER: Not necessarily. There might be, somewhere in the universe, black holes of relative low mass, say, a mass of one billion tons, commensurate with the mass of Mount Everest.

NEWTON: How do you want to produce that? It cannot be done by stellar collapse.

HALLER: You are right. But there might be turbulent events in the universe that have such holes as by-products. That could have happened in the early days of the universe, which we will discuss shortly. In contrast to the stellar and galactic black holes that formed during the cosmic evolution, we have a name for black holes that have been around from the beginning; we call them *primary black holes.*

There is the interesting fact that holes with a mass of a billion tons have a lifetime of an order of magnitude of 10 billion years. And that is what we believe to be the age of the universe. And that again means they should be around today as radiating objects, possibly even as exploding ones.

NEWTON: You mean that such radiating holes are moving freely around the universe and might even come close to Earth?

HALLER: That would depend on their density. And that we can only guess at. We might suppose that many primary holes were created in the Big Bang, where things were really turbulent. But there is no chance that too many such holes might be in our universe today. Black holes emit intensive gamma radiation—energetic photons—as they explode. And these gamma rays could be seen. Now, we do see gamma rays, but we can explain their existence by normal processes, without taking recourse to black holes. And that means that primary black holes of masses in the range of a billion tons or less cannot be a common feature. An upper limit for their medium density is 300 in a cosmic cube one light year across. Nevertheless, the density of these holes could be much larger inside our galaxy—gravity would see to it that black holes are concentrated inside the galaxies, just like stars.

EINSTEIN: Therefore, would it be useful to look for exploding black holes in the universe?

HALLER: Absolutely, and we do that when we look for what we call gamma ray bursts—showers of gamma rays that pop up all of a sudden. But, to date, we have not seen any events that give clear indications of black-hole explosions. Still, we cannot exclude that sometime in the near future we may find first evidence for the existence of black holes in this fashion.

But the time has advanced to a late hour. So, let's leave the black holes alone and turn to more concrete matters. I took your assent for granted and reserved a table for dinner in the Panda Restaurant. That is a Chinese establishment in the north of Pasadena. And tomorrow we'll meet again here in this library.

CHAPTER 18

The Cosmic Beat

We are just an inconspicous part of Nature: we feel
how little a single creature means, and it makes us
happy.

—Albert Einstein[1] (entry into his diary
during a sojourn on the English coast)

The next morning, right after breakfast, the meeting continued in
the Athenaeum library. Again it was Newton who opened the discussion: "If we put an electric charge somewhere into space all of a
sudden—and that is easily done with today's technology—that
charge will be surrounded by an electric field of force. And that will
happen with the speed of light, as we discussed when we talked
about the special theory of relativity.

"Now I ask myself: what happens when, all of a sudden, I place a
heavy mass here in this room? How that can be done technically is
of no importance—this is just a gedanken experiment. The effect of
this mass on the surrounding space—its gravitation—is tied to the
laws of the theory of relativity, just like the effect of an electrical
charge. That means that the gravitational effect of the mass does not
immediately exist all over space; rather, it will take a short time. If
there were a gravitational field, I would say it builds up with the
velocity of light. But since such a field does not really exist, and since
gravitation is a consequence of the structure of spacetime, we have
also to conclude that the change in spacetime structure—this per-
turbation of the fabric of spacetime with the sudden appearance of
that mass—happens with the speed of light. In other words, there is
something we might call gravitation waves, that is, vibrations of the
structure of spacetime."

EINSTEIN: You're not telling me anything new here. These vibrations of our spacetime structure, which happen with the speed of
light all across the universe in the form of gravitational waves, are a

direct consequence of my gravitational field equations. In this regard, my equations resemble those in electrodynamics which were formulated by your compatriot James Clark Maxwell in the middle of the nineteenth century. Maxwell's equations provide the theoretical framework of electrical engineering to this day.

NEWTON: I don't know the details—but as far as I know, electrical charges do not radiate off any energy as long as they are at rest or in uniform straight-line motion. Only when they are accelerated will they radiate. For example, electrons that swing back and forth—that are being periodically accelerated and decelerated—generate the radio waves emitted by a radio antenna. Can we say the same for gravitational waves? Can we build up a transmitter of gravitational waves by moving large masses back and forth?

EINSTEIN: Of course. An object at rest or in constant straight-line motion will not generate oscillations in the fabric of spacetime. Only when it is accelerated will there be an emission of gravitational waves. When I push down the gas lever in my car, it will emit gravitational waves. But please don't ask me to prove that by experiment. Such gravitational waves are much too weak ever to become observable.

NEWTON: That's an interesting point. It puts a new twist on the phenomenon of inertia. An object that, due to its inertia, resists any attempt to change its state of motion, will also resist the emission of gravitational waves. It doesn't want to transmit them—and that's why we call it inert.

EINSTEIN: You said it. But let me add that this does not look like an explanation of the phenomenon of inertia, if that is what you had in mind.

NEWTON: I'm not sure myself—but it is certainly well worth pondering. For right now, let's discuss energy: electromagnetic waves do transmit energy. Is that true for gravitational waves also?

EINSTEIN: Of course it is. Let me cite a small example. Let's imagine we have two metallic spheres; one of them has a positive electric charge, the other a negative one. Let them orbit each other. Now, during this motion, the electric field around them will change all the time, and that will generate electromagnetic waves of the same frequency as that of the orbital motion. These waves will take some amount of energy out of the rotating system—an effect somewhat similar to friction. The rotating motion will slow down and will finally stop, putting the spheres to rest.

NEWTON: I got it. If I consider two masses out in space that orbit each other due to their gravitational interaction—say, a two-star system—they will radiate off gravitational waves: and these waves will remove energy from the two-star system. If we had ways of observing such a system over a sufficiently long time, we would be able, at least in principle, to detect this radiating-off of energy. But, of course, not by direct observation. Indirectly, we would observe the loss of energy. That would not be a definitive proof of the existence of gravitational waves—but it would be a step in the right direction. It would be important to know how strong the effect is.

HALLER: The quantity of energy that is radiated off by these gravitational waves can be calculated on the basis of Einstein's equations. There is no need to worry about the details here. But it is instructive to look at a few examples. In this context, imagine a device we could easily construct here on Earth: imagine an iron rod 100 meters long, weighing 1,000 tons. Let's put this rod on a rotational device so that its ends move in circular motion with a radius of 50 meters. We make it rotate at a velocity that is just below the level where the material would be pulled apart by centrifugal forces—and let that be three revolutions per second. The energy loss by the emission of gravitational waves will be 10^{-26} watts per second.

NEWTON: That is a forgettably tiny amount, isn't it?

HALLER: There are two reasons for this. For one thing, the coupling of matter to the fabric of spacetime by Einstein's equations is proportional to Newton's gravitational constant G, and that constant is quite small. Also, the radiation of gravitational waves, just like that of electromagnetic waves, is dictated by the velocity of light. What counts here is the relation of the velocity of the endpoints of the rotating rod as it compares with the speed of light. If that ratio were 1:2, about 200 kilowatts would be radiated off, and that is quite a lot. Within one hour, our system would emit 200 kilowatt hours of power.

NEWTON: Does that mean we have to look at very rapidly rotating systems if we want to measure gravitational radiation?

HALLER: That would certainly help. Energy emission from a two-star system can be quite impressive—10^{20} watts or more. I will give you a useful example. Take two neutron stars. Each has twice the solar mass, and they rotate around each other at a distance of just 100 kilometers, and with a frequency of 100 revolutions per second.

That makes their velocity about 30,000 kilometers per second, or one tenth of the velocity of light. The energy radiated off per second will, in this case, be approximately 10^{45} watts.

There are particularly impressive emissions of gravitational radiation when we have catastrophic happenings of cosmic dimensions such as happens, say, with the collapse of a massive star into a neutron star, accompanied by the explosion of a supernova. Another example would be the melting together of two neutron stars, which can also lead to a black hole. Both of these processes take a very short time only, about one thousandth of a second. We can estimate that the explosion of a supernova will emit an amount of energy on the order of 10^{33} to 10^{36} kilowatt hours with its gravitational waves. When two neutron stars fuse, a similar amount of energy will be radiated off—actually a bit more, of the order of 10^{38} kilowatt hours. We are talking about amounts of energy large enough to see them as commensurate with fractions of the rest energy of the Sun, according to Einstein's mass-energy relation. When we have supernova explosions, we are talking about something between one millionth and one thousandth of the rest energy of our Sun that is being radiated off.

EINSTEIN: I'd like to make you aware of one important fact here. The gravitational radiation, when emitted by the kind of cosmic catastrophes that we've mentioned, originates in the very region where the catastrophe happens, in contrast to optical signals. A supernova explosion, on the other hand, is visible all right; but the sudden appearance of such powerful light emission originates from the surface of the exploding star—and not from its interior, which might give us much more interesting information. If we could observe gravitational waves just as we see light waves, we would be able to learn a lot about what exactly happens during that catastrophic event.

HALLER: That is correct. But there is another aspect: gravitational waves, in contrast to electromagnetic waves, will not be disturbed or distorted by surrounding matter, even in large quantities. Spacetime oscillations propagate undeterred across entire galaxies, and across the most distant parts of the Earth. If we had telescopes that observe by way of gravitational waves instead of light waves, the cosmos would look quite different. Light waves show us nothing but the surfaces of the stars. Gravitational waves, on the other hand, give direct information about the essential dynamics of cosmic processes. The structure of the universe revealed by gravitational effects would be much more spectacular than what is revealed to us by light.

NEWTON: So you are saying that the fabric of our spacetime is constantly being jolted by the gravitational thunder that comes from cosmic catastrophes. We have no devices that would permit us to change gravitational waves into sound waves—wouldn't I love to listen to the resulting cosmic concert, if we did!

HALLER: I would join you as a listener right away. But we'll soon see that there might be hope, that we may be able to listen to such a concert in the foreseeable future. Imagine the melting-together of two neutron stars into a black hole. This would happen over a time span as small as 0.01 second, and our two neutron stars would be orbiting each other with an angular velocity of one thousand revolutions per second. The energy that would be radiated off then corresponds to one half of the rest energy of the Sun. This energy results in a brief warping of the structure of spacetime, and this warping propagates with the speed of light from the location of the catastrophe. It is a great deal larger than the energy radiated off by our Sun—by a factor of 10^{21}.

EINSTEIN: Our galaxy consists of about 100 billion stars. The energy you mentioned would then correspond to the power radiated off by 10 billion galaxies—and that means by about 10 percent of all the galaxies in the universe visible to us. Isn't that incredible? And we are talking about nothing more than the blending together of two neutron stars.

HALLER: But let me stress that this energy is contained in a spherical shockwave, with a width given by the time of the melting together of the two stars. We said that this time is 0.01 second. The width of a shockwave then amounts to 0.01 second times c, the velocity of light. The resulting spherical shell is then 3,000 kilometers thick, as it propagates across space.

EINSTEIN: I just did a little exercise. I made an estimate about the strength we would feel on Earth if such a blending process happened in the center of our Milky Way. I am amazed by my answer: we are talking about 100,000 watts per square meter. That is 100 times the power radiated onto the surface of the Earth by our Sun's light. And it also means that spacetime is being shaken quite badly when a black hole is formed by the blending of two stars in the center of the Milky Way. Even on Earth, that originates quite a thunder. But I can imagine a happening much more violent than the blending of two neutron stars. I could think of the blending of two black holes. That really would amount to a cosmic orgy. If it happened in our galaxy, it would cause the structure of spacetime to be warped quite badly even close to Earth. But of course, we have to ask ourselves how often such catastrophes could happen in our galaxy.

HALLER: Your estimates certainly show us that it makes sense to look for gravitational waves on Earth. Getting back to the energy emitted by a system consisting of two neutron stars—we mentioned how large the amount of energy is that such a system radiates off by means of gravitational waves. This effect has actually been observed for one such system, code-named PSR 1913 + 16. One of those two neutron stars happens to be a pulsar, hence the designation PSR in its name—and it emits radio waves at regular intervals, which help us to follow the motions in this system fairly precisely: the two neutron stars orbit each other once every eight hours. They approach each other along spiral paths as the energy is radiated off, and the velocity with which they approach each other gives us a measure of the energy that is being radiated.

EINSTEIN: In other words: the size of the effect can be calculated precisely in the framework of my theory. I guess I have to be worried that this means a real test for it.

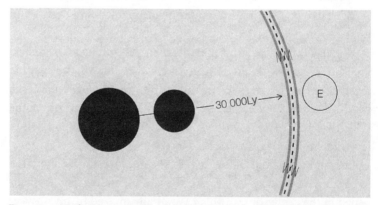

FIGURE 18.1 When two neutron stars fuse together to make a black hole, they radiate off energy in the form of a gravitational shockwave some 3,000 km thick. When an event of this sort occurs close to the center of our galaxy, the power of its radiation that reaches the Earth is 100 times larger than that of all solar energy impinging on the surface of the Earth.

HALLER: I wouldn't worry, Mr. Einstein. It has already been shown that the amount of energy lost agrees with the general theory of relativity, and that it does so with astonishing precision. The theory predicts that one orbit will change within a year by the tiny fraction of 2.7 parts in a billion—and that is precisely what was observed. Your theory passed the test with high marks.

NEWTON: Let me congratulate you on that, my dear colleague. I can see that your theory has essentially no chance to be proven wrong. And still, this does not do more than provide an indirect hint that there may be gravitational waves. But weren't we looking for a direct proof of their existence? Let's give this one a try. Let's look at a gravitational wave that issues from some cosmic catastrophe—say, from Einstein's orgy of the joining of two black holes. The wave will propagate through our Earth with the velocity of light. Hence, there will be brief distortions in the structure of spacetime. But what exactly will happen? And how will we see it?

HALLER: Since a wave such as this one would have traveled across a great deal of space before arriving at the location where we observe it, we are justified in looking at it as a plane wave—a wave that propagates through space looking like an actual plane at any given moment. Examples that we know are the sound waves originated by a distant detonation or the growling of a distant thunderstorm. Let's look at this scenario in a coordinate space where we choose the plane of the wave as the xy-plane. The wave propagates at right angles to this plane, in the direction of the z-coordinate. It leads to a kind of oscillation of the spacetime metric, similar to oscillations we observe with metal plates. These oscillations of the metric lead to rhythmic distortions of space, changing from dilation in the x-direction while compressing the y-direction to the inverse of this order.

The intensity of this wave is marked by the relative strength of the distortion, i.e., by the ratio of the dilation of the length divided by the length itself. We have no trouble estimating what size the effect will be on Earth for the case of those waves that are generated by the blending of two neutron stars in the center of our galaxy—which would amount to no more than 10^{-18}.

EINSTEIN: That is painfully small. If I think in terms of a kilometer's length, this kilometer will be changed by no more than 10^{-13} centimeters. And that's about as much as the diameter of an atomic nucleus, which means it is practically immeasurable.

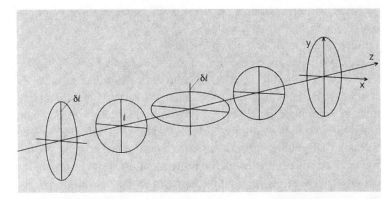

FIGURE 18.2 Effects of a plane gravitational wave: there is a succession of space dilation and space compression, which we describe by the variations $\delta\ell$.

HALLER: That is what we would normally think. But, in reality, the case is not hopeless. As long ago as the 1960s, our colleagues began to think about ways that could help us pin down the existence of gravitational waves. The idea was that we observe with great precision reverberations of a big metallic container, say, an aluminum cylinder that weighs a ton. When a gravitational wave impinges on a cylinder like this one, it has an effect like the hammer with which the percussionist in an orchestra hits a xylophone. One of the cylinders of the xylophone will be excited and oscillate in its eigen-frequency—the frequency that is fixed by its mechanical structure. In effect, the wave excites sound.

The distortion of the fabric of spacetime, then, is what causes the oscillation of the cylinder. The problem with a detector of this kind is this: how does it become sensitive to very small oscillations, such that we can notice them? What's more, we have to make sure that the excitations we observe are actually due to a gravitational wave, and not to some background effect. To reduce such backgrounds from external influences, it makes sense to build two different detectors— in the slang of the field, they are called cylindric antennae—at a fair distance from each other. Our modern technology permits us to measure distortions of the metric of a relative size of 10^{-18} with the aid of such metallic cylinders.

EINSTEIN: That is truly impressive. That does imply that today's cylindric antennae should be able to observe a cosmic catastrophe such as a supernova, when it occurs in our galaxy.

HALLER: Sure enough—but too bad that nobody knows when we'll have the next supernova explode in our galaxy! The statistics of past occurrences do not look very promising. There have been observations of supernovae in the Milky Way in the years 1054, 1572, and 1604; and ever since, all has been quiet on the galactic supernova front. In 1987 a supernova explosion was registered to have occurred in the Great Magellan Cloud—a stellar system in our immediate vicinity. It has been determined that our present-day detectors would have been able just barely to see a signal of the gravitational waves this explosion must have emitted. Unfortunately, it turns out that just at the time it occurred, all detectors were switched off for technical reasons—and so we really missed our chance.

EINSTEIN: It is clearly risky to wait for a supernova explosion somewhere close by; so we would do well to improve on the sensitivity of our detectors. That might put us in a position to observe supernova explosions in other, neighboring galaxies; and it would make sure that something observable pops up every year, more or less.

HALLER: This is the way present research is laid out. The most recent detectors are trying for a sensitivity of 10^{-21}—and that makes them about a thousand times more sensitive than the cylindric antennae we mentioned.

NEWTON: We really should be in a position to measure the distortions of the spacetime metric over larger distances.

HALLER: And that is precisely the direction that was taken. We now are no longer talking about cylinders a few meters long; we are now talking contraptions that measure several kilometers.

EINSTEIN: Gravitational waves, as we know, foreshorten space in one direction while expanding it in the direction perpendicular to it. So one might think of a new version of Albert Michelson's famous experiment. You recall that this experiment compared the velocities of two light pulses that propagate along lines perpendicular to each other. Today we know that, to satisfy the special theory of relativity, these two velocities have to be the same. If, now, a gravitational wave propagates through our device, there would have to be a tiny difference in timing between the two directions, since the lengths contract or expand, respectively. And if we take advantage of interference phenomena, we should be able to be quite sensitive to such effects.

HALLER: What you just proposed is the principle of a gravitational interferometer. It works similarly to the Michelson experiment, but it is much more sensitive; it uses laser technology that was not available to Michelson. Today we are designing such interferometers with lengths for the two channels that conduct the laser light that extend to several kilometers. And this is the way in which we hope to raise the sensitivity to the value of 10^{-21}, as mentioned before.

Here at Caltech, a prototype of such an interferometer with an "arm length" of 40 meters has been built; another one with lengths of 30 meters was constructed in Munich, Germany. The principal investigator of the local gravitational wave project happens to be a friend of mine; so you'll forgive me if I have already arranged with him that we take a look at the Caltech interferometer this afternoon. I hope that meets with your approval.

EINSTEIN: But of course we'll take a look at that. I would love to see how those laser beams operate in this environment.

HALLER: The array we will look at is just a prototype for the large detector now being constructed. Its acronym is LIGO, which stands for Laser Interferometer Gravitational Wave Observatory.

LIGO consists of two separate interferometers, and its arm-

FIGURE 18.3 The interferometer at Caltech, with an arm-length of 40 meters. The laser beams are propelled inside the two long vacuum tubes that are at right angles to each other. (*Photo:* California Institute of Technology, Pasadena)

FIGURE 18.4 Model of one of the two LIGO interferometers, this one being built close to Hanford, Washington. (*Photo:* California Institute of Technology, Pasadena)

lengths measure 4 kilometers each. One of the detectors is positioned close to Hanford, Washington, and the other one is just outside Livingston, Louisiana. The detectors, as you see, are built apart at a great relative distance. In this fashion, background effects that have nothing to do with gravitational waves, such as local tremors, can be eliminated.

In Europe, physicists are constructing detectors close to Pisa, Italy, and in the vicinity of Hannover, Germany. The arm-lengths of those instruments are 3 kilometers each. In addition, there is talk about more such projects in Australia and in Japan. We might wind up with a whole network of interferometers.

The detection of gravitational waves with the help of several different interferometers should be able to give us details about these waves. Suppose a tremor in the spacetime metric caused by a happening in a distant galaxy might finally reach our solar system: it might be seen at first by the detector at Hanford, then by the one in Louisiana, and finally by those detectors in Europe. We could then analyze time differences between these various signals to find out which direction the waves took. In addition, the precise wave form and its frequency would give us more information about the origin of the gravitational catastrophe—be it a supernova, the collapse of a system of neutron stars into a black hole, or the fusion of two black holes into a single one—each of these processes produces gravitational waves of a characteristic kind and magnitude—each of them, you might say, has its own gravitational fingerprint. And that means that this new century is seeing the appearance of a new discipline in astronomy opened up by detectors of gravitational waves. Who knows, sometime in the distant future it may become equivalent to optical astronomy, or to radio/X-ray astronomy.

EINSTEIN: Let's stress the fact that, to date, no gravitational wave has been observed, much less the frequencies and wave forms of such spacetime waves. But you may well be right. Gravitational waves might open up a whole new window looking out into our universe. Through this window, we might have a chance to get a different glimpse at cosmic happenings, to gain new insights into the depth of spacetime, to take a peek into the pits of cosmic catastrophes.

CHAPTER 19

The Dynamic Universe and a Blunder of Einstein's

All my efforts take it for granted that our world is
based on a completely harmonic structure.

—Albert Einstein[1]

In the morning after the visit to the LIGO group, our threesome reconvened in the dining room of the Athenaeum. Einstein told his two colleagues about his plan for a visit to the Mount Wilson Observatory: "When I came to Pasadena in 1929, it was there that Edwin Hubble confronted me with facts that dramatically confirmed the cosmological ideas that had been derived by means of my equations years before. That is why I suggest we should start our discussion of cosmology on Mount Wilson today. I'm sure that you, Haller, know the observatory from previous visits; but it will be interesting for Newton to see the very place where quantitative cosmology was born in the late 1920s."

Right after breakfast, they took the ride up toward Mount Wilson in Haller's rental car. They drove north on Hill Street to route 210 of the freeway. From there, they quickly reached Angeles Crest Highway, winding their way up into the San Gabriel Mountains. Finally, a road veered off toward Mount Wilson, and a little side street took them to the large parking area next to the observatory.

It was a clear day. An eastern wind came off the Mojave Desert, doing Los Angeles a favor by blowing its smog out over the ocean. Toward the south, the metropolis filled the large visible basin. In the distance, the skyscrapers along Santa Monica beach lined the horizon, stretching into the blue ribbon of the Pacific, interrupted only by Catalina Island.

Haller pointed out to his colleagues the white dome on the distant mountain to the southwest, Mount Wilson observatory's big

FIGURE 19.1 Partial view of the Caltech campus, with the San Gabriel Mountains as a backdrop. The tall building on the right is the Millikan Library. (*Photo:* California Institute of Technology, Pasadena)

FIGURE 19.2 Albert Einstein together with astronomers of the Mount Wilson Observatory in front of the building housing the large mirror telescope, around 1930. (*Photo:* Hale Observatory, Pasadena)

brother, located on Mount Palomar. After World War II, Mount Palomar's mirror telescope, also erected by Caltech astronomers, would become the largest such telescope in the world. Next, there was a visit to Mount Wilson's small museum, documenting a collection of old instruments and also displaying photos of Einstein's previous visits to Mount Wilson. Some showed Einstein with Edwin Hubble and his closest collaborator, Milton Humason.

They finally came to the big circular building that houses the telescope. A side entry gave them access to a vestibule, from where they could see inside the observatory's dome structure.

Einstein was voluble: "When I came here for the first time, Hubble showed me how the big telescope is operated. That was, unfortunately, in bright daylight; clearly, there was no way to get a peek at

FIGURE 19.3 The 2.5-meter mirror telescope in the Mount Wilson Observatory. (*Photo:* Hale Observatory, Pasadena)

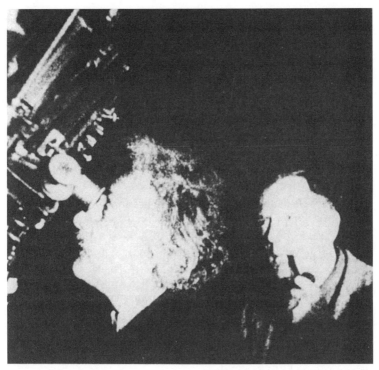

FIGURE 19.4 Albert Einstein and Edwin Hubble at the mirror telescope of the Mount Wilson Observatory in the early 1930s. (*Photo:* California Institute of Technology, Pasadena)

the stars. While Hubble explained the apparatus, I sat on this little old wooden chair which, I am happy to see, is still here. From that vantage point, Hubble looked out into the universe farther than any man had done before him."

Einstein pointed to an old chair standing on a small platform— and sure enough, it was marked as Hubble's chair; it had been left there to honor the great astronomer.

HALLER (*to Newton*): Construction of this observatory began early in the twentieth century. In those days, the sky above Southern California was clear and free of smog. The region around Los Angeles was considered very appropriate for observational astronomy. There was frequently an inversion layer of the atmosphere that caused a stabilization of the air strata—an Eldorado for astronomers, who were happy to take advantage of the favorable climate to pursue their nocturnal work without being exposed to cold air.

The big mirror telescope which, at the time of its construction, held the record with a diameter of 100 inches, or 254 centimeters, was designed by George Ellery Hale and financed largely by the Carnegie Foundation. Edwin Hubble, who had been a Rhodes scholar in Oxford as a young man, started his astronomy research on Mount Wilson right after the World War I.

Right away, Hubble's first efforts set new milestones for research in astronomy. He was able to prove, based on data from the new telescope, that assumptions made by several astronomers in the past— and even by the philosopher Immanuel Kant—had been true: the distant nebulae are indeed stellar configurations just like our own Milky Way. This meant that the universe open to observation increased in size by many orders of magnitude.

EINSTEIN: We should call Hubble the Copernicus of the twentieth century. It was Copernicus who moved the center of the universe from the Earth to the Sun. Later astronomers then realized that the Sun, in turn, is nothing but an average star of medium size inside a large galaxy. It was Hubble who then realized that our Milky Way is an ordinary and nonspectacular galaxy in a vast cosmos—just one galaxy among billions of others.

NEWTON: For Hubble to realize that the spiral nebulae are distant stellar systems, he had to estimate their distance from Earth. How did he go about that?

HALLER: His new telescope permitted Hubble to look at individual stars inside galaxies that are close to us. Among these were stars with a periodically changing brightness—the so-called Cepheids. By comparing their luminosity with other stars in our own galaxy, he was then able to estimate their distance.

NEWTON: The distance from our Sun to the center to our galaxy is about 30,000 light years. How far is our distance from the next galaxy?

HALLER: On Earth, we measure distances in units of kilometers or miles. Inside our galaxy we are usually dealing with thousands of light years—the diameter of the Milky Way is of the order of 100,000 light years. But if we are interested in intergalactic travel, an appropriate unit of length is a million light years, corresponding to 9.46 times 10^{18} kilometers. Our distance from the next large galaxy, which belongs to the Andromeda configuration, measures about 2 million light years.

NEWTON: I'm trying to come up with a limit that tells me about the size of our visible universe, such as it was determined in the wake of Hubble's discovery. What are the largest distances that have been identified in our universe?

HALLER: It is obviously difficult to come up with a precise number, but the appropriate order of magnitude is 10 billion light years, or 10,000 million light years.

NEWTON: All right—let's suppose we redefine a million light years to equal one meter. That would rescale the entire universe that can be seen through the telescope as a sphere with a radius of 10 kilometers; in its center, there would be our galaxy—an accumulation of stars with a diameter of 10 centimeters, about as big as a grapefruit. The Andromeda galaxy would be at a distance of 2 meters. What do we know about the distribution of galaxies in space? Do they move around in some arbitrary place or pattern? Or do they form distinct structures?

HALLER: Hubble, in his time, was not in a position to say anything about the distribution of galaxies in space. That became possible only in the 1960s and 1970s with the help of observational methods. Today, we know that there are many galaxies located in what we call galactic clusters. First hints of their existence also came from Caltech. They were identified by Hubble's colleague Fritz Zwicky. The closest galactic cluster is located in the direction of the constellation Virgo, about 60 million light years away from us, or 60 meters on our new cosmic-length scale. The Coma cluster is at a distance of about 300 meters, in the direction of a configuration called Coma Berenices. The universe is replete with such galactic clusters, with diameters that measure 100 million light years or more. It is astonishing that there are also gigantic holes in the universe—giant spaces that appear to be empty, that hold not a single galaxy. These holes may well measure 100 meters on our cosmic scale.

EINSTEIN: To make a long story short, I suppose I can picture the universe that we see through telescopes as a sphere with a 10-kilometer radius; inside, we have the galaxies, disk-shaped configurations about 10 centimeters in diameter; they in turn are grouped together in clusters measuring some hundred meters. And then there are these big empty holes, about the size of the galactic clusters. Being a Swiss citizen, I would say the cosmos looks a bit like Swiss cheese. Half cheese, half air. I might now ask: "Whence this strange structure?"—but that would take us away from the main point of our discussion.

HALLER: You are right about that. We will come back to your question later—it does indeed have a bearing on the early history of our cosmos. But before that, I'd like to talk about Hubble's second great discovery.

EINSTEIN: Just a minute—wouldn't it be better if we did things in the same sequence as chosen by history? And that means we shouldn't start with Hubble on Mount Wilson but rather with the Einsteins on Haberland Street in Berlin, Germany.

But I am getting uncomfortable in this cool room—why don't we continue our discussion in the warm California sun outside?

Close by the telescope building, they resumed their conversation in a comfortable shady nook.

NEWTON: You really got me curious about the connection between Haberland Street and the early history of the universe.

EINSTEIN: I wish that connection was even closer than it actually is, but the first steps were done there, in the development I will describe. And were it not for a fatal error I committed, the second step would have been a sure thing.

Shortly after writing down my gravitational equations, I started applying them to the cosmos as a whole. I made two assumptions: for one thing, I took it to have, on average, a homogeneous structure; the matter density should be the same everywhere, when averaged out. This is, of course, an idealization. But it should be good enough for a first try.

HALLER: In the final analysis, your assumption turned out to be pretty good. We now know that the matter density in the universe is fairly constant once we average over large distances—distances that have to be large enough to comprise entire galactic clusters.

EINSTEIN: Too bad my second assumption did not turn out to be that successful. There, I assumed that our cosmos is, ultimately, a static structure—that it is the same today as it was a long time ago, and that it will essentially remain this way forever.

NEWTON: I think there is a problem. A long time ago, I determined that a universe that consists of many massive objects which attract each other gravitationally cannot be static—for a very simple reason: every massive object attracts every other massive object. As a consequence, there has to be an implosion in the long run—matter has to collapse. A static universe cannot be stable.

EINSTEIN: I was certainly aware of that. As I included the two assumptions into my equations, the result that you inferred from your own theory did turn up: the cosmos cannot be static, it has to collapse. In other words, I was not able to find a static solution for my equations.

Finally, I came up with a ruse: I modified my equations by adding a small term formulated such that it was compatible with the principle of general relativity without, however, adding to gravitational effects. In this way, this term could not be excluded due to gravitational measurements; but it could still be important over cosmological distances. This added term was assigned the role of a force that counteracted normal gravitation; we might call it an antigravitational force; or, better, a kind of cosmic pressure that is built into the fabric of spacetime, and that would push matter apart if there weren't normal gravity.

NEWTON: I get it. You are arranging a balance between gravity, which wants to have our cosmic home collapse, and cosmic pressure, which drives it apart.

HALLER: We have big tennis halls these days with walls and roofs that are airtight. A pump sees to it that there is always a slight amount of extra air pressure inside the building, which is stabilized in this fashion. Without this extra pressure (which corresponds to Einstein's term that he added to his equation), the tennis hall would finally collapse due to gravitation.

This example also shows that this cosmological term is a new quantity and, as such, a parameter in our theory, just like the gravitational constant before. But it cannot be chosen arbitrarily; it has to match the matter density such that there will be a balance between gravity and this pressure that opposes it.

EINSTEIN: Be that as it may—I have to admit I made a mistake when I invented this cosmic "pump." I have often had the occasion, in later times, to admit that my invention of what has become known as the *cosmological constant* was the greatest scientific blunder I committed in my entire life. I wish I had never thought of it.

HALLER: I can agree with you only partially. We will soon see that what you call your greatest "blunder" has recently come back into serious discussion—if from very different quarters.

EINSTEIN: Well, I will certainly look forward to finding out the details once we get there—but you will pardon my skepticism.

Now back to the year 1916. With my new cosmological constant, I had constructed a static model of the universe; it looked like a three-dimensional version of an inflated balloon. Just like the surface of the balloon, space was uniformly curved and of finite size. The counterpressure of the cosmological constant kept the universe in equilibrium, and for a while I liked what I had done.

NEWTON: Now tell me more about the stability of your static model universe. I'm not sure your analogy with the balloon makes sense to me. The balloon is a static structure as long as I can keep its pressure constant. A universe replete with massive objects that attract each other, on the other hand, is a very fragile structure. Your cosmological constant will stabilize the system only if the matter inside is completely homogeneous in its distribution. But I would expect gravitational effects to cause considerable fluctuations of the density here and there—as in the case where a number of stars congregate accidentally in some particular region of space. That might well stir up instabilities, and problems for your equilibrium.

EINSTEIN: Well, I'll admit that you have caught on to the Achilles heel of my model. It does not, as you say, lead to stability. Bring in the density fluctuation, and there goes our cosmic equilibrium. In other words, my model universe turned out *not* to be a model for our actual universe.

I wish I had had the courage, at the time, to play around with my equations. But the patience that was needed belonged to somebody else—to **Alexander Friedmann** in the Soviet Union. Friedmann assumed, as I had done, that the matter density is the same across the cosmos, but no more than that. For this scenario, my equations will turn into an equation for the density of matter. Friedmann found, for starters, that my equations lead inexorably to a change of matter density with time. In other words, there was no solution that gave a static universe. Imagine a rock that is thrown upward, only to drop back down after a while, due to gravitation. The rock cannot remain at rest—it has to move because of prevailing effects of gravity. Newton's law of gravity permits us to calculate the path taken by the rock: for any given time, we can determine the height of its path, if only we know its initial velocity. In the same way, Friedmann was able to calculate the density at any given time, assuming we know the density at one specific point in time.

NEWTON: That means Friedmann differed from you when he found out that matter density is not constant with time—that it is likely to increase or decrease.

EINSTEIN: It is not only the density that will change—that also applies to the curvature of space. Take the universe, in the framework of my first model, to be a closed space, the three-dimensional analogue of a spherical surface. That will lead to two possibilities: either space will be inflated like a balloon, or it will get smaller. For the self-inflating space, its matter density will keep decreasing. The volume, of course, will not keep expanding forever—it will reach some particular maximum size. And after that it will recontract.

Our example with the rock is useful also for this configuration. A rock that is thrown upward will reach its maximum height at some point and will then fall down again; note that I am excluding the case where its initial velocity is great enough so that it could escape the gravitational pull of our Earth.

NEWTON: So we might imagine our universe in terms of a finite large space that is closed and curved, and uniformly populated by galaxies. This space increases, it expands; the galaxies move apart; but the gravitational forces between the galaxies see to it that the expansion gets slower and slower, and that it finally stops. There must be gravitational contraction after that, making galaxies move back together. Now I'm asking myself what happens when their relative distances are so small that they actually permeate each other.

EINSTEIN: Now this is a discussion I would prefer not to go into right now. Recall that I did not maintain that our universe could actually be described by one of Friedmann's models; I just described various possibilities.

HALLER: Let me stress that the uniform drifting apart of the galaxies is an effect of the expansion of the space they populate. In principle, the galaxies are at rest—each one has an assigned location in space. But the space between the galaxies is inflated; an astronomer living in one of the galaxies would have to observe that all other galaxies move away from his, and with a velocity that is the larger the farther removed they are from one another. Our astronomer gets the impression that he is sitting in the center of the universe, while all other galaxies are moving away from him.

Another astronomer in another galaxy would have exactly the same impression; every point in space, every galaxy, is accorded the same rights. This is cosmic democracy, just as in the case of a balloon: once we inflate it, every point on its surface will move away from every other one. There is no central point that could be considered as the origin of the expansion.

EINSTEIN: Let me mention another example in three dimensions: let's take a yeast cake made up of dough that has a homogeneous mix of raisins, and bake it in an oven. The yeast pushes the dough apart in the baking process, and the raisins move along with the dough. The expansion of the cake makes the raisins follow the pattern of the galaxies in Friedmann's model of the expanding universe. The greater the distance between any two raisins, the greater their relative velocity. Every raisin is at rest with respect to the dough in its vicinity, it is embedded in the dough.

NEWTON: Sure enough, the dough is equivalent to intergalactic space, but here it is the yeast that drives the dough apart; what is the medium that does the same for space? Is there a cosmic analogue for

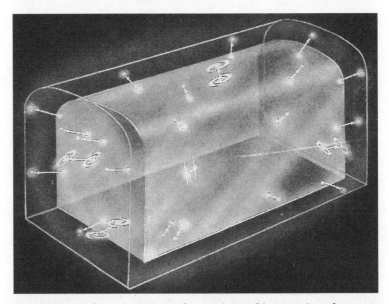

FIGURE 19.5 As the universe expands, we picture this expansion of space as being analogous to that of the dough when baking a yeast cake. (Graphics by Wendlinger, *P.M. Magazine*)

FIGURE 19.6 Edwin Hubble at the mirror telescope at Mount Wilson Observatory, sitting on his legendary wooden chair. (*Photo:* Hale Observatory, Pasadena)

yeast? In your Berlin talk on the matter, Haller, you mentioned the complicated structure of the vacuum; so maybe there is such a phenomenon.

HALLER: Friedmann does not need cosmic yeast for the expansion in his model. Once expansion is realized at any given point in cosmic time, it will continue—the principle of inertia is valid in the cosmos and applies to cosmic matter. Cosmic expansion cannot stop all of a sudden.

EINSTEIN: That gives me only half an answer to my question. You did not tell me why there is expansion in the first place.

HALLER: That is a point I don't wish to touch on just now—we will come back to it in any case later.

NEWTON: Fine, but I do have another question. We are talking about an expansion in space. Space is increasing in size all the time—sort of like a rubber band that we keep stretching. All distances increase with time. But what about the massive objects in space—what about stars, planets, the rock I'm holding in my hand? Do they increase in size? I suppose they don't, because we would not have any reason to talk about expansion in the first place if the scale that we use to measure length expanded at the same rate.

HALLER: You are right again. Cosmic expansion of space, such as in Friedmann's model, implies that the cosmic scales as measured by the distances between galaxies will change in comparison with earthbound scales—say, the size of this rock, or the diameter of an atom. We do observe that the size of a rock cannot change continuously with time. That follows from the fact that matter is made up of atoms. Every atom has a given size. If we line up 10 billion hydrogen atoms, one next to the other, we will have a line one meter long. The cosmic expansion of space will not change that: the laws of quantum physics determine the size of the atom. Cosmic expansion then means an expansion of space relative to the size of the atom. Space as such increases, the atoms remain the same—just as in our cake. There, the dough inflates, the raisins don't.

NEWTON: You were mentioning the rock that gets thrown upward, and then will fall back. Now we know that it will not drop back to Earth if its initial velocity is larger than about 11 kilometers per second. Is there a chance that the cosmic expansion in Friedmann's model could be so fast that there will not be any subsequent contraction? That, just like the rock vanishes into the universe, the universe itself could keep inflating forever?

EINSTEIN: You are anticipating what I meant to get to next. It really depends on the velocity of expansion, whether there will be a subsequent contraction. If the expansion proceeds sufficiently rapidly, it will slow down in the course of time, but it will never stop. Whether that happens or not depends on the relevant matter density—on the number and mass of galaxies per cosmic unit volume. If there is enough matter, the expansion will finally stop; if not, it will continue forever.

One interesting detail is Friedmann's discovery that the decisive question—"Will expansion continue forever?"—is closely tied to the structure of space. There will always be a final contraction for space that is itself closed in, like the three-dimensional analogue of a spherical surface that we discussed. In this case, matter must be so densely arranged that the curvature caused by this density will close the space, as we say. That leads to a universe of a finite size, a closed universe. The opposite occurs when matter is less densely packed. This is the case that corresponds to a saddle surface. You will recall that this one has negative curvature, and is infinite in size.

NEWTON: So in this case, the universe would be infinite, but still uniformly filled with galaxies. And that means there would be infinitely many galaxies.

EINSTEIN: Yes indeed- the universe is infinitely large and will keep expanding forever.

Let me mention an interesting limiting case: assume the expansion to be just barely rapid enough so it will continue forever. This corresponds to the case where the rock thrown upward has just barely enough energy to vanish into space but will come to rest eventually. That will happen when its initial velocity as it leaves the surface of the Earth is 11.2 kilometers per second.

NEWTON: That marks the fine line between positive and negative curvature of space. What does Friedmann's theory have to say about the curvature in this particular case?

EINSTEIN: This is where things are really simple. There is no curvature: space is planar, Euclidean—just as we like to imagine it. And it makes sense, since there cannot be either positive or negative curvature. So it must be zero. But still there is expansion. Space is flat, planar, without curvature, but it keeps increasing in size; and like any Euclidean space, it reaches out to infinity.

HALLER: So let me summarize: when we apply Friedmann's equations—which are, after all, just your gravitational equations—to the universe as a whole, and if masses are homogeneously distributed there, we have two cases:

1. The universe is closed: space has positive curvature and has finite size. The curvature is determined by the matter density.
2. The universe is open: the curvature is negative. Space reaches out to infinity; its matter density is not sufficient to lead to a closure of space.

The limiting case between the two prevails if the matter density is just such as to stop the expansion of space in the distant future. There is no curvature of space.

All models share one feature: there is no static universe. Space either expands or contracts—the cosmos has its very own dynamics.

NEWTON: Now that is enough about Friedmann's model. After all, it's nothing but a model solution of Einstein's gravitational equations. Whether it is relevant for our Nature, for our own cosmos, I have no way of judging. But am I wrong in guessing that you would not have gone into so much detail unless Friedmann's model were close to what describes our universe?

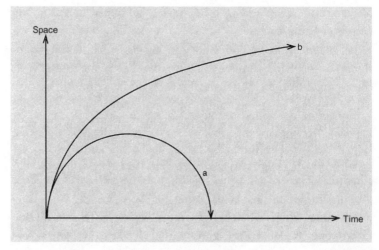

FIGURE 19.7 The two possible versions of cosmic dynamics: (a) in a closed universe, a period of expansion is followed by one of contraction; (b) in an open universe, the expansion is permanent.

HALLER: That brings us to the point where our thoughts should revert to Edwin Hubble. But time has passed quickly, and so we should attack the lunch problem first. I was prudent enough to have the Athenaeum kitchen's crew prepare a picnic basket. We are now close to Charlton Flats; it is a lofty place without visible curvature, but with a beautiful pine forest; let's have a leisurely alfresco lunch there.

CHAPTER 20

A Discovery on Mount Wilson

To me, the search for discovery is one of these independent pursuits without which a thinking human would not know how to affirm his right to life.

—Albert Einstein[1]

It took only half an hour before the picnic on Charlton Flats was in full swing. Haller had chosen a spot under a large pine tree that afforded an excellent view of the observatory's outer buildings.

Haller was well acquainted with the region. Years ago, the annual outings of the Caltech physics department used to wind up with an ample picnic right here. On one of these occasions, he had engaged Stephen Hawking in a ball game. At the time, Hawking was spending a year at Caltech—and that was shortly after he had discovered that black holes radiate energy. He was already relegated to his wheelchair; but he nonetheless managed to move about in it with agility.

It took little time before our three physicists were relaxed and in the best of moods; Haller had brought along a bottle of a fine California Chablis, and that certainly helped—not even Newton could resist it.

"I can see you are getting a bit impatient," Einstein started out. "You want to find out which puzzle Edwin Hubble managed to solve during his nightly escapades up here. Fair enough, I will give you a short version of what happened.

"During the 1920s, Hubble was investigating the light from distant stars. He used spectral analysis—that is, the investigation of the distribution of different colors of light emitted by the stars—to confirm that the stars in neighboring galaxies contain the same chemical elements as our Milky Way—hydrogen, helium, lithium, and so on. Their atoms, when at great heat, emit photons of characteristic energies that are well known to us from experiments on Earth.

"Now, when we look closely at the light that is emitted by, say, hydrogen in a distant galaxy, we can find out whether the photons of which it consists have the same energy as those that are measured on Earth in similar experiments. The trick is that, of course, this energy will change when that galaxy is in motion relative to our Earth. If the galaxy we are looking at moves away from Earth, the photon energy decreases a bit; conversely, it will increase when that galaxy approaches our Earth. In other words, the motion of that galaxy subtracts or adds a bit of energy to those photons.

"You might compare it to what happens to a gun aboard some ship. If it fires a shot while that ship is in motion, the energy of the projectile will increase when the ship is moving toward the aim of the shot; it will decrease when the motion goes in the opposite direction."

NEWTON: All right, so you can find out the relative velocity of a distant galaxy from an analysis of the light it emits. And what was it that Hubble found?

EINSTEIN: When I came to visit him here on Mount Wilson, he showed me the distribution of velocities for quite a few galaxies. They had a number of different values, typically amounting to thousands of kilometers per second.

NEWTON: And I suppose that some of them moved in the direction of the Earth, while others moved away.

EINSTEIN: Strangely enough, that is not what he found. The first astonishing result was this: the more distant galaxies are all moving away from us, the faster the farther they are from us.

HALLER: This is the effect we call the redshift of the light from distant galaxies. As they move away from Earth, the light they emit loses energy; and that means its wavelength increases. In the process, the color of this starlight moves toward the red—because the energies of the photons of red light are smaller than those of blue light. That is the effect which closely resembles what each of us has noticed when a police car moves by us: when that car moves away from the observer, the sound of the siren has a lower pitch than when the car is at rest.

NEWTON: So a redshift gives us a hint of cosmic expansion, as in Friedmann's model?

EINSTEIN: Patience, please! Hubble had no idea of the theoretical implications but, intuitively, he did the right thing—he tried to find out whether there was a relation between the motion of the galaxies and their distance from us. True enough, it was a difficult task to determine those distances with some precision, and it turned out later on that he was not able to avoid a few systematic errors while doing that work. But that is not the point here. The first thing Hubble noticed was this: galaxies that are roughly at the same distance from us share the velocities at which they move away from the Milky Way. This was a first hint that there is a dependence of velocity on distance. He then drew up a diagram of those velocities versus their distances from Earth. It became clear that these velocities increase with the distance—their flight velocities are proportional to the distances, and that is exactly the effect expected in the framework of Friedmann's model, in analogy to what we said about the raisins in the yeast cake as that cake expands in the baking process.

When Hubble showed me his data, I was convinced that he was right: the universe is expanding. That means that, as early as 1915, my equations of the general theory of relativity had given me the key for an understanding of cosmic dynamics—I just had not made any use of that. And so it was left to Friedmann to find the correct solution, although he never realized that his "ansatz" correctly described the dynamics of our universe; the poor fellow did not live to have that satisfaction—he died at age thirty-seven. Without knowing about Friedmann's work, the Belgian astrophysicist Abbé Georges Lemaître derived Friedmann's model again in 1927. He stressed, above all, the simple relationship between the flight velocities of the galaxies and their distances from us, which Hubble had noticed.

On that day, I experienced an odd preoccupation: I had derived the equations of gravitation from simple principles and on the basis of everyday observations, making use of no more sophisticated tools than pen and paper. And there I was in front of that mighty telescope, noticing that my equations, simple products of our thought processes, were actually able to describe the dynamics of our entire universe. I had the feeling that Hubble and I were profiting from a unique opportunity to look over the shoulders of the Old Man upstairs, getting a peek at the blueprint of our universe.

NEWTON: I do envy you, my friend. It is a miracle: we cook up theories and concepts by pondering the principles Nature reveals to us; and here we are, finding that, in the process, we have been able to write down a description of cosmic dynamics not just in our solar system but far away in the most distant regions of the universe. Matter may be tremendously diverse in the ways it appears, but the basic architecture of the universe must ultimately show a great simplicity that is more or less the same for the entire cosmos.

EINSTEIN: It is the simple things that are so hard to accomplish. This is what Bertolt Brecht once told me when I talked to him about my work in Berlin, a long time ago.

NEWTON: So Hubble was able to determine the rate of space expansion from his measurement of the flight velocities. And what is that rate? Suppose I look at some given distance in intergalactic space, and let's assume that distance equals the 1,000 kilometers that we travel from London to Berlin—by how much will it change in a year's time?

HALLER: To date, we don't have an exact value, and the precision with which we know it may have a 20 percent margin of error; but I can give you an order of magnitude: every cosmic distance increases by about 10 percent in a billion years. The galaxies that are moving apart today must therefore have been located in close proximity some 10 billion years ago. And that means that the order of magnitude of the time during which the universe has expanded is about 10 billion years. We can call that the age of our universe—and I should stress that we are talking about an order of magnitude here, not a precise age. That may be anywhere in the range between 12 and 16 billion years.

So, a cosmic distance will increase every year by a fraction 10^{-10}. And for the distance between London and Berlin, we are talking about a rate of expansion of 0.1 mm per year. The rate of expansion of our universe as we measure it today can be stated in terms of kilometers per second for every million light years, or for each **parsec**— and one parsec equals 3.26 light years. The rate that is accepted by most astronomers is about 20 kilometers per second per million light years, or 65 kilometers/second per parsec. That means the endpoints of a stretch of space a million light years long are moving apart with a velocity of 20 kilometers per second. Larger distances, of course, move apart at larger velocities. A galaxy that is a hundred million light years from Earth then moves away with a velocity of 2,000 kilometers per second (100 x 20 = 2,000 km/sec).

FIGURE 20.1 The measurements of Edwin Hubble and Milton Humason showed that distant galaxies—such as the spiral galaxy shown here, which resembles our own Milky Way—are moving away from us. This is a consequence of the expansion of space between our galaxy and this spiral galaxy.

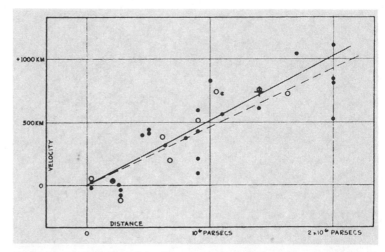

FIGURE 20.2 A 1929 plot of the dependence of the flight velocity of galaxies on relative distances, discovered by Hubble.

EINSTEIN: As long as we are dealing with small distances, Hubble's law of expansion does not apply: the velocity of expansion is subject to interferences from other velocities of local motion. The Andromeda galaxy—the one that is closest to our Milky Way—is at a distance of some 2 million light years. Hubble's law would make it move away from Earth at a rate of about 40 km/s. But, in reality, it actually approaches the Milky Way, a consequence of the gravitational pull between these two galaxies.

HALLER: The same is true for the galaxies that belong to large galactic clusters. Inside such a cluster, there is no such thing as cosmic expansion: gravitation dominates the dynamics of relative motion inside the cluster. But when we are dealing with distances larger than about 50 million light years, uniform expansion is the prevalent effect. This phenomenon is generally accepted by now.

I should really mention to you that there is a fairly wild story behind establishing the rate of expansion. The value that Hubble first determined implied a cosmic age of about 2 billion years. Now we do know that our Earth is older than that, so this value is not acceptable. The universe cannot really be younger than its progeny, among which we have to count our Earth. Then, in the 1950s, Hubble's colleagues Walter Baade and **Allan Sandage** figured out that all of Hubble's conclusions about cosmic distances had to be revised

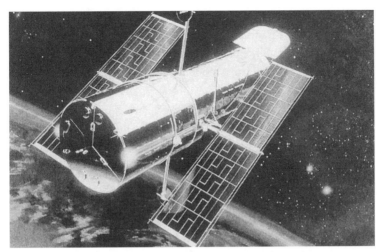

FIGURE 20.3 The Hubble space telescope is a mirror telescope stationed in space, about 610 km above the surface of the Earth. It permits, among other things, a great increase in the precision with which we can measure distances between the Earth and faraway galaxies. (*Photo:* NASA)

upward by a factor of 10. An important step in the right direction was due to Sandage's measurements that he did with the help of the Mount Palomar Observatory, which had been erected on that peak in 1948. But to this day, the precise rate of expansion is not really known to an uncertainty better than a factor of about 1.5. We do expect that this situation will soon change, when we start getting data from the new Hubble space telescope. In all likelihood, we should soon be in a position to determine the rate of expansion to within 10 percent of its true value.

EINSTEIN: Now, when you do speak of the rate of expansion, you are referring to its value at the present time, which means some 12 to 15 billion years after its start. But we know that, according to Friedmann's model, this expansion decreases with time. The rate of expansion is not some fixed number, it changes with time. In principle, we might measure this slowing of the rate of expansion when determining the redshift of distant galaxies; note that when we observe a galaxy today, what we actually see is that galaxy as it was a long time ago—at the time it emitted the light now entering our instruments today. And at that time, it had a larger velocity. Has this effect of cosmic deceleration been observed?

HALLER: You make a good point there; but the deceleration is measurable in the manner you describe only when we also know the distance from which the light reaches us. And we are aware that our knowledge in this regard is not so hot. At any rate, there has not been any proof of the deceleration of expansion to this day. If it decreases at all, it is most probably by a slow rate—which is what Friedmann's model makes us expect. As a result, our astronomers are not at all alarmed that we have not made any positive observation of this effect to date.

NEWTON: If we did know the amount of deceleration, that would tell us about the density of matter in the universe: that is the quantity which, by means of the ensuing gravitation, determines the slowing down of receding galaxies. But we could tackle the question differently. We might ask how much matter we need to get a closed universe: how much matter do we need so that our cosmos finds itself on the brink between a closed and an open universe?

HALLER: In the limiting case, space would be Euclidean; it would have no curvature. The exact value of the median density of matter depends on the expansion rate. Let us assume, for a minute, this rate amounts to 20 km/s for every million light years. That would tell us that the matter density we need for an expanding universe without curvature amounts to 8 times $10^{-30} g/cm^3$. And that is equivalent to about 5 hydrogen atoms per cubic meter. This is what we call the critical mass density. If the cosmic mass density is smaller than that, the universe is open; if it is greater, the universe, in contrast, is closed in itself.

EINSTEIN: And what do we know today about the mass density that actually prevails in the universe?

HALLER: The ratio of mass density and critical density is usually denoted by the Greek letter omega. If we distributed all luminous matter—all the light-emitting stars and gas clouds—evenly in space, we would remain far short of the critical density. We would arrive at only a fraction of that value. Omega has a value of the order of a few percent.

NEWTON: That means it appears that our cosmos is open, that space reaches out to infinity. Too bad omega has a pretty small value: I would have been much happier about a value of omega = 1, about a value that has the limiting amount.

EINSTEIN: I can just about imagine why you like that. It would imply that space has no curvature; space would look the same as in Newton's mechanics. But, Sir Isaac, it does not look like this is what Nature has chosen. Haller just told us that the data favor an open universe—a universe with negative curvature.

HALLER: Don't rush us, dear Einstein. I mentioned that there is not enough *luminous* matter to effect a closure of space. But we cannot exclude the possibility that there are new forms of matter that are not like the light-emitting stellar matter; there might well be something we call "dark matter," which is recognizable only by the gravitational pull it exerts. We have already come to know an example for that— black holes. But we can currently estimate that black holes formed from the collapse of stellar matter do not contribute much to the over-all mass density. At least, it is not enough to make omega = 1.

EINSTEIN: We have also already talked about Fritz Zwicky's dis-covering signs for nonluminous but gravitationally effective matter in galactic clusters—what he called missing matter. Is there a chance that this matter, whatever it turns out to be, can contribute enough mass to make omega = 1?

HALLER: As far as we can tell today, the dark matter that we can-not identify or describe amounts to about ten times the mass of nor-mal stellar matter.

NEWTON: That is really unbelievable! It means that the gravita-tional dynamics of galaxies and galactic clusters is completely dom-inated by dark matter. Whatever matter we see in the shape of stars is nothing more than a little encore for the clouds of dark matter— sort of the seasoning in the soup, one might say, which also makes up most of the taste but nonetheless adds almost nothing to the sub-stance.

HALLER: We have had certain hints that dark matter is not based on atoms but rather on some different objects—such as neutrinos, for instance. Recall that I mentioned these neutral relatives of the electrons in my Berlin lecture. There also may be completely novel particles we have not discussed—there is no boundary for what par-ticle physicists cook up!

The search for what constitutes dark matter is in full swing—it is one of the most pressing problems in today's physics and astro-physics. The fraction that dark matter contributes to the mean den-sity of matter in space determines whether the expansion of the cosmos will continue forever, or whether the universe is infinite rather than finite.

EINSTEIN: That does sound ironical: indeed, it sounds like a new Copernican upheaval. You will recall that Copernicus was the one who found out that our world is not part of an array centered by the Earth, but that it looks heliocentric instead. It took a while before astronomers then found out that Giordano Bruno was correct when he said that the Sun is only one of a hundred billion stars in our Milky Way. And then came Hubble, who downgraded our galaxy to the rank of an enlisted man in a whole army of galaxies. And now we are finding out that most of the mass in the universe consists of a component that has little or nothing to do with normal matter—the matter that makes up the stars, the planets, ourselves—the matter built up out of atomic nuclei and electrons. If this is indeed the case, we have all the more reason for modesty, given that we exist in the form of a kind of matter that makes up only a minuscule minority of the true cosmic mass.

NEWTON: Never mind—I would vote in favor of a mass density omega factor that equals 1. That would permit us to forget the curvature of space. This value of one, I think, presents the most reasonable picture: if omega were a bit smaller than 1, we would have to ask ourselves why this is so and not a whole lot smaller still. It is astonishing enough that we find omega hovering in the vicinity of 1—yet it is not one millionth, or ten thousand. That cannot be pure coincidence. I will argue in favor of our taking a closer look at the case where omega $= 1$.

HALLER: I promise you, Sir Isaac, that we will soon get to this point. But I think we need a little physical exercise now, don't you think? Would you be amenable to a stroll through the woods while we continue our conversation?

CHAPTER 21

Echo of the Big Bang

All I really want is to sit back and find out how God created this world. It is His thought that I am trying to understand—not the spectral lines of this or that element. I really could not care less about things like those.

—Albert Einstein[1]

Haller recalled there was a path that crossed the pine forest just below the crest of the San Gabriel Mountains. Our threesome quickly decided to take a longer walk. It did not take long before the discussion resumed when Newton turned to Haller: "At the end of your Berlin lecture, you mentioned the Big Bang. At the time, I considered that to be a speculation made up out of thin air. In the meantime, we have been discussing cosmic expansion, and that made me reconsider. Could we really conclude, just by observing that the galaxies have been rapidly moving apart ever since they were closer together some 10 to 15 billion years ago, that the universe actually started its existence at some precise time, and that there was a cosmic explosion in the beginning?"

EINSTEIN: We have seen that the collapse of a massive star leads to a black hole—to a singularity in spacetime. Let us imagine we have a cameraman who makes a documentary film of a collapse of this sort. If, later on, he projects the film backwards, what will he see on the screen?

NEWTON: He will notice that, in the beginning, there is a state of high matter density, which then expands rapidly. The black hole becomes undone, and finally vanishes. And in the end we see the star just as it looked before the collapse.

EINSTEIN: This process clearly resembles what we discussed as cosmic expansion. Now let's assume we watch a movie of cosmic dynamics as it runs backward. Expansion now becomes contraction: the galaxies approach each other; they start overlapping, so that, eventually, stars from different galaxies melt together. At the very "end," we will wind up with a dense soup of nuclei and electrons—something we call a cosmic plasma.

HALLER: Thus, at the end of our movie, which would of course be the beginning of our universe, we may have a singularity similar to a black hole—a virtual state of infinitely high density and temperature.

NEWTON: You mean you really are serious about the notion that our cosmos might have originated in a singularity?

HALLER: At the very least, that is a good possibility. The singularity would be the beginning of time, of space, and of matter. It would make no sense to speak of time before that Big Bang—just as it is meaningless to talk about a temperature below absolute zero.

EINSTEIN: At any rate, the equations on cosmic dynamics that Friedmann developed tell us that, in the scenario of an expanding universe, there must have been a state of infinitely high density and temperature in the beginning. Now, it might be that the laws of quantum physics see to it that there are no true infinities, just as in the case of black holes: singularities pop up in mathematics, but Nature does not like them. But that is not the point here. What we want to retain is this: in its beginning, our cosmos was not only very densely packed, it presumably also had a very high temperature. In that state, it was similar to a grenade just after it explodes. Cosmic expansion is a direct consequence of the power of that explosion.

NEWTON: And where do we get an indication of that high initial temperature?

EINSTEIN: I don't really know this; but I surmise it is true, in analogy to a collapsing star, where a part of the kinetic energy of collapsing matter gets transformed into heat energy.

HALLER: Today, we know that the early universe did have a very high temperature—from the phenomenon of cosmic thermal radiation. To this day, we can observe the electromagnetic radiation that was emitted in the Big Bang.

NEWTON: You must be joking. The Big Bang, if it happened at all, dates back more than 10 billion years. Any thermal radiation that may have been emitted at that time would surely have vanished by now.

EINSTEIN: Easy, Sir Isaac. Do you recall Haller's argument in his Berlin lecture? If the cosmos was once very hot, it was full of thermal radiation. It has been expanding ever since, and so has the thermal radiation that was part of it. As a consequence, the immense space of the universe, including huge volumes of intergalactic space, are not really empty; they are full of photons. These photons cannot vanish: photons in empty space are conserved—they can neither be created nor annihilated.

HALLER: In my lecture I mentioned that, on average, we have some 400 photons per cubic centimeter today. That makes photons the dominant particles in the universe. But to call them thermal radiation is essentially a joke. Due to all that expansion, the radiation that measured many millions of degrees in the beginning has cooled down to a temperature of no more than 2.7 degrees Kelvin. We should really call it cold radiation, if it were not for the fact that physics considers all matter to have some degree of warmth, if not heat in the common sense of the word, as long as its temperature is above absolute zero.

NEWTON: Now, did you tell us that this radiation has been found experimentally? How so? Is it not impossible to do experiments in faraway intergalactic space?

HALLER: True enough, but that is not needed. The discovery was made on Earth, in the eastern part of the United States—in New Jersey, to be precise. The radio astronomers **Arno Penzias** and **Robert Wilson** did experiments with a large microwave antenna, sensitive to radio waves of a wavelength in the range of about 7 centimeters; its purpose was the reception of signals from the telecommunication satellites Echo I and Telstar.

In 1965, while making these observations, they accidentally noticed an odd background radiation—a kind of humming that hits the Earth uniformly from all directions in the skies. And this radiation looks just like the thermal radiation with some 3 degrees Kelvin.

EINSTEIN: If that really is thermal radiation, it will have to satisfy the law derived by my good friend Max Planck, who was my colleague in Berlin in the beginning of the twentieth century. It says that the wavelength of this radiation is not fixed; rather, it has a characteristic distribution, with the maximum of its intensity distribution depending on the temperature.

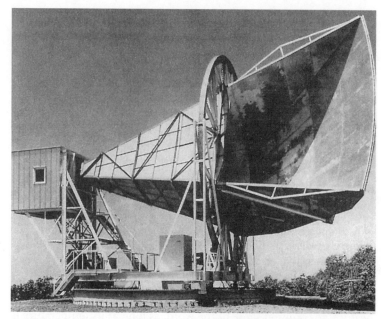

FIGURE 21.1 The horn-shaped antenna, located in Holmdel, New Jersey, of Penzias and Wilson, which enabled them to detect cosmic background radiation in 1965. (*Photo:* Bell Telephone Laboratories)

HALLER: In the 1980s, a satellite was put into space with the explicit purpose of measuring the wavelengths of this cosmic thermal radiation. This satellite, with the acronym COBE, for Cosmic Background Explorer, was a great success: it turned out to do more than anybody could have expected.

The intensity distribution that COBE measured as a function of the wavelength of this radiation turned out to reproduce Planck's theoretical expectation closely. What is particularly astonishing is the isotropy of the radiation (in plain English, this means it comes from all directions uniformly). The temperature with which it hits us when coming from the constellation Ursus Major is just the same as that which we get from the direction of the Southern Cross.

EINSTEIN: That means the 2.7 degrees Kelvin radiation is some kind of electromagnetic echo of the original explosion, a radiative leftover of the Big Bang. And that makes two things even more likely: first, that there *was* a Big Bang to begin with and, second, that the temperature prevailing at the time was extremely high. Actually,

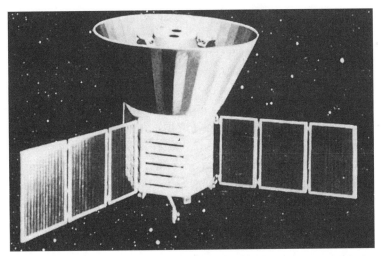

FIGURE 21.2 The COBE satellite, which measured cosmic electromagnetic radiation. (*Photo:* NASA)

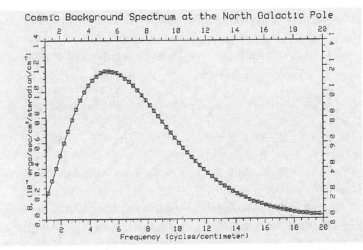

FIGURE 21.3 Wavelength distribution of cosmic electromagnetic radiation as measured by the COBE satellite. The entire line represents the theoretical distribution according to Planck's law of radiation, for a temperature of 2.7 degrees Kelvin.

I think one could have derived a prediction of this radiation from these two hypotheses.

HALLER: Its discovery was more or less accidental. The two people who found it did not even suspect there might be some cosmic radiation of this kind. But the effect had actually been predicted in 1948 by George Gamow and his collaborators Ralph Alpher and Robert Herman. In their article in *Physical Review*, they speak of a background radiation with a temperature of about 5 degrees Kelvin. But it looks like their work was not taken very seriously by experimental physicists—had that not been so, the phenomenon should have been discovered some ten years earlier. But we should not forget to mention that this prediction on the part of Gamow and his collaborators also included other predictions on the origin of the elements which later turned out to be wrong.

As far as the precise temperature is concerned, it could not be predicted; it depends on the density of normal matter, among other things. It makes sense to compare the number of photons per unit volume with the number of nuclei. On average, we find that the ratio of the number of photons per cubic centimeter dwarfs the number of nuclei by a factor of the order of 10^{10}, or 10 billion: there is only one nucleon per 10 billion photons.

EINSTEIN: If the universe was all that hot in the beginning, there must have been a time when matter was a hot soup made up of protons and neutrons—and this soup cooled down rapidly. I would think that, in the beginning, the number of neutrons and protons was about equal: the nuclear forces between these particles do not know which is which. Would we not expect that, in today's universe, one should have as many protons as neutrons bound in the atomic nuclei?

HALLER: No, but you are moving in the right direction. A short time after the Big Bang, there must have been an equal number of protons and neutrons, because of the prevailing symmetry. Now, when the proton-neutron porridge cools down, the ratio of protons to neutrons will change: the mass of the neutron is a bit larger than that of the proton, and that will have its effect when the temperature of the universe, expressed in energy units, is comparable to the relative mass difference, given in units of energy.

This opens the door to processes that change neutrons into protons, reactions similar to radioactive neutron decay, while the opposite reaction can no longer occur—the proton has a lower mass than the neutron. About three minutes after the Big Bang, when the temperature was just moving past 900 million degrees, the nucleon soup has only 13 percent neutrons, but 87 percent protons.

As the cosmos keeps cooling down, further atomic nuclei will be formed. Before long, free neutrons will no longer be around. They are mostly bound up in helium nuclei which, we know, have two protons and two neutrons each. It does not take much computational skill to figure out that, about half an hour after the Big Bang, the nuclear matter in the universe consisted of helium and protons. At that time, helium made up 25 percent of the mass total, and protons made up the rest. The "dough" from which stars and planets were to be formed, later on, was mostly hydrogen and helium.

EINSTEIN: Great! The Big Bang formed a lot of helium. Is that an explanation for the fact that the stars have not only large amounts of hydrogen but also a lot of helium?

HALLER: Yes, precisely. There is no other mechanism by which such a large helium fraction of the cosmic mass density would have formed during stellar evolution. When we look at stars that are young enough so as not to have had time to produce other atomic nuclei in nuclear processes, we find that helium is, next to hydrogen, the main building material.

EINSTEIN: Well, we have wound up foursquare in cosmology! That was a field wide open for all kinds of wild speculations when I had my first serious discussion on the genesis of the universe in the early 1930s with Hubble. At that time, there was not much known about the dynamics of atomic nuclei and about all those elementary particles—all aspects of physics that are of an obvious importance for the first moments just after the Big Bang.

But at present we don't have the time to delve into the details of particle physics. I would suggest we now look into the first moments after the Big Bang, even though I am aware that much of what we can discuss must remain fairly speculative.

By this time, the three hikers had reached a rocky promontory east of Mount Wilson, which afforded a beautiful panoramic view toward the south and southeast. In the distance were the San Jacinto Mountains, and to their right the luminous peak of Mount Palomar and the desert hills of Anza Borrego. They tried to find comfortable seats on the rocks, while Adrian Haller started to give a little lecture about physics just after the Big Bang.

HALLER: I would like to divide the discussion about cosmology into two parts. In the first, I would like to talk about what we get when we apply our present-day knowledge of particle physics to cosmology. It will turn out that this does not suffice for an understanding of a few important aspects of our cosmos; this forces us to resort to theoretical extrapolations that have not yet been proven by experiment. That's what we will discuss in the second half; I guess we will get to that point either tonight or tomorrow morning.

We know by now that the microcosm of elementary particles is governed by two different kinds of forces. For one, there is a strong force between the quarks which is, as we mentioned before, transmitted by gluons. This is the same force that dominates inside atomic nuclei just as it acts between the nucleons. The gluonic forces have the interesting property that they do not become weaker when we deal with large distances, such as happens with electric attraction. This means that the quarks cannot be observed as free particles: they can be seen only indirectly, such as in certain scattering experiments. Three quarks are bound together in one nucleon.

NEWTON: I have no trouble understanding that two objects will bind together, such as a proton and an electron when they are pulled together by electric attraction, to then form a hydrogen atom. But why three quarks? Why does it take three of them to form one proton and not just two?

HALLER: I was waiting for that question. It has kept physicists busy for years: it was known from indirect evidence and, later on, from direct experiments, that a proton contains three quarks. But the nature of that force which constrains three quarks to form a tightly bound proton remained a mystery until, ultimately, it was discovered that quarks have a completely new property that had never been seen in other particles. To wit: quarks come in three different incarnations, a bit like water that may show up in the form of ice, water, or vapor. The three incarnations are completely equivalent. Since there are just three of them, they were tagged in analogy to the three basic colors that television uses for the construction of color pictures; red, green, and blue. So, we speak of the three "colors" of quarks—and, mind you, this is merely an analogy because the properties of quarks have nothing whatever to do with real colors.

This color property rather resembles the electric charge in electrodynamics—with the difference that there are three color "charges," but only one electric charge. That is why we need three quarks, each of which has a different color charge, to make up a proton or a neutron. The nucleons then are, as we say, color neutral: they don't have a net color charge; the colors of the three quarks blend together and neutralize each other—just as we know it in optics, where the superposition of red, green, or blue colors combines to make up white, a color-neutral state. In this sense, we can call the proton and all the other nuclear particles either "white" or colorless.

Normal stellar matter, such as the matter on our planet, consists of two different types of quarks, which we tag u and d quarks, where u and d simply stand for *up* and *down*. Nucleons are be built up out of these quarks; the proton is made up as a (uud) state, the neutron as (ddu). But there are additional types of quarks that come in pairs like u and d—four more of them, to be precise. The second pair is made up of the *charm* quark, code-named c, and the *strange* quark, code-named s. The third and last pair contains the *top* quark we already discussed, and the *bottom* quark: these two are called t and b, for short. And, needless to say, all these quarks have the above-mentioned property that they carry the color charge.

EINSTEIN: I may not know the details, but those quarks certainly look to me like the elementary constituents of matter. I do think those particle physicists might have come up with a little more imagination when they gave those quarks their name: top and bottom, up and down—the fundamental particles of Nature, I think, should have deserved better names.

HALLER: You may well be right, but I'm afraid we are too late in the game to change any of these names. But we'll do as most physicists are doing, making use only of the initials.

NEWTON: And what, if I may ask, makes the new quarks different from u and d?

HALLER: Mostly their mass. The masses of the quarks that make up the nucleon, of the u and d, are very small; c and s are considerably heavier. But the heaviest of the quarks is, as we said before, the t-quark. This one is the heaviest subatomic object that has ever been seen; its companion, the b-quark, has a mass that is 35 times smaller—it has about 5 GeV.

The u and the d quarks are the only ones that make up stable particles—such as the proton or stable atomic nuclei. All particles that contain one of the "new" quarks are unstable; they decay right after their creation in particle collisions, into other particles! The only quarks that are present at the end are u and d.

The strong force between the quarks is mediated by gluons, as we said. Just as photons act on the electric charges to generate electrical forces, the strong nuclear force between quarks is generated when the gluons act on the quarks' color charges—we might call the mechanism a kind of color dynamics. In fact, physicists call this color dynamics of the quarks and gluons **chromodynamics**. But it is not only the gluonic forces that act on quarks; there are also the weak nuclear and the electromagnetic forces. It is the weak force that causes radioactive decay, such as the decay of the neutron—but also the decay of the heavy quarks. It is mediated by the W and the Z particles. And, of course, we recall that the electromagnetic force is carried by Einstein's photons.

NEWTON: Now you add the weak and the electromagnetic forces to the gluonic one. You are now talking about three forces, while only a short while ago it was just two.

HALLER: And there is a reason. It turns out that the weak and the electromagnetic forces are two forms of one and the same force. The difference that we see in experiments when dealing with these two kinds of force is due to the fact that the carriers of the weak nuclear force are quite massive—that, in fact, they have a large mass—but that the photon is massless. This difference is irrelevant at very high energies—as for instance in the time shortly after the Big Bang. And I should mention that it is caused by the same field that is responsible for the generation of masses.

EINSTEIN: So now you are talking about the mysterious Higgs field again, which you mentioned in your lecture as the field that describes the properties of our vacuum.

HALLER: It is by means of the interaction of the W and Z particles with the Higgs field that these particles acquire their masses; the photon does not do likewise. Again, it is the structure of the vacuum that is responsible for an important phenomenon—for the difference of the strong and electromagnetic forces. I need not tell you how important that is for the way in which Nature presents her macroscopic image. If the W and Z particles were massless like photons, the world would look totally different. It is the mass of the W and Z that makes the weak forces weak, considerably weaker than the electromagnetic ones.

EINSTEIN: Nevertheless, I will insist that this unification stands on clay feet: nobody, as we discussed recently in Berlin, actually knows whether the Higgs field does or does not exist.

HALLER: To unify the weak and electromagnetic forces we don't necessarily need the Higgs field. It is entirely possible that the masses come about some other way. But that would presumably not do anything to the basic idea of unifying the two forces. Their unity does not consist in a vague, artificial tying together of these forces. The theory predicts that the carrier particles of the forces—the Z and W bosons on one hand, and the photons on the other—interact in a very specific fashion with each other. This prediction has been tested in great detail at the LEP accelerator in Geneva.

But now let me add a few words on an additional class of particles, on the leptons, the best known of which is the electron. Its name—and for once it has a serious meaning—is derived from the Greek adjective *leptos*, which means light: the electron, after all, has a mass that is a great deal lighter than the proton's. Unfortunately, this well-intentioned name turned out to be less useful when, in the 1970s, a heavy cousin of the electron was found: the tau-lepton, which has a mass twice that of the proton.

The leptons, like the quarks, appear in pairs. And as in the quark case, there are three different pairs: there are three leptons with a negative electric charge—the electron, the muon, and the tau-lepton, sometimes also called the tauon. Each of these has an electrically neutral partner, a neutrino. Their lack of an electric charge does not permit neutrinos to participate in the electromagnetic interaction. It is only by means of the weak nuclear force that neutrinos can interact with other particles.

EINSTEIN: I recall that Wolfgang Pauli postulated the existence of a neutrino particle, totally unknown at the time, in the late 1920s. He did that to explain some inconsistencies in neutron decay. I once talked to Pauli about this particle of his. I told him, a bit ironically, that he had invented a ghost particle, since nobody could really prove its existence. He actually agreed with me. Pauli assumed there would not be any way to observe neutrinos directly, due to the very weakness of their interaction with other matter. He was unhappy about that: physicists, unlike philosophers, are not pleased when they invent something that must remain unobservable.

HALLER: Pauli would have changed his mind today. He was quite mistaken. True: their very weak interaction does not make it easy for us to observe them. A neutrino can traverse the entire Earth, our entire galaxy, without any trouble. Only extremely rarely will it have a reaction with an atomic nucleus or with an electron. That tiny rate will increase with the energy of the neutrino. Today, the CERN laboratory in Geneva and the Fermi National Accelerator Laboratory outside Chicago have facilities with intense neutrino beams that permit sophisticated experiments. Pauli, when he made his theoretical arguments some seventy years ago, could not have foreseen that we can now generate a lot of neutrinos with high energies that make direct observation possible, if only once in a while. But even so, most of the neutrinos being generated in particle collisions in these big laboratories manage to get away quite easily. The CERN neutrino beam penetrates the Jura mountain range close to CERN, and then takes off from Earth in a straight line trajectory. Its neutrinos become cosmic vagabonds forever.

Some 170,000 years ago a supernova exploded in the Great Magellan Cloud—a cluster which accompanies our Milky Way on its path across the universe. During such stellar explosions a large percentage of the stellar mass is transformed to energy, and this energy then gets emitted in the form of electromagnetic or neutrino radiation. At that time, a shockwave consisting of photons and neutrinos was propagated with the velocity of light across interstellar space; it hit the Earth on February 22, 1987. Astronomers immediately observed this explosion. Later it was estimated that at precisely 11:35 P.M. California time, many billions of neutrinos traversed the Earth. A few of them were actually detected by means of two large detectors that had been designed for completely different purposes—one in Japan, the other in the United States. That was the beginning of an entirely new branch of astronomy, which we now call neutrino astronomy.

FIGURE 21.4 Partial view of the Japanese underground detector Kamiokande-II, located in an old zinc mine. We see the large photomultiplier tubes that detect photons from particle reactions. In 1987 they helped register charged particles from eleven neutrino interactions in water. The neutrinos originated in the 1987 supernova detected 170,000 years after its explosion in the Great Magellan Cloud. Our picture shows a section of the detector's volume before it was filled with water.

Certainly, neutrinos are no longer the ghost particles they were in Pauli's time. With today's techniques, there are several ways in which we can detect them either directly or indirectly.

NEWTON: And you said there are three and only three neutrinos; how do we know that?

HALLER: It has been quite a while since it became clear that there cannot be many different kinds of neutrinos in the universe. Let me spare you the detailed arguments. But the actual test for the 3-neutrino hypothesis was performed at the LEP accelerator at CERN: it turns out that a precise observation of the decays of Z particles will tell us when there are 2, 3, or maybe even 4 different kinds of neutrinos. The more different kinds of neutrinos that exist, the faster the Z particle can decay. And there was a quick and direct answer: there are three species of neutrinos. Our universe does like the number three.

EINSTEIN: When we talked about dark matter in the universe, you mentioned that neutrinos might have a mass. Tell me more about that.

HALLER: For one thing—considering the physics of the Big Bang—there are good reasons to suppose that the number of neutrinos per cubic centimeter, all over the universe, is of the same order of magnitude as the number of photons.

NEWTON: But we know there are 10 billion times more photons than nucleons. And now you claim that there are also 10 billion times more neutrinos than nucleons?

HALLER: This is what we infer from the hot Big Bang: a mechanism similar to the one that generates the ocean of photons in intergalactic space will also lead to the formation of an ocean of neutrinos—about 450 neutrinos per cubic centimeter, 150 each for the three varieties of neutrinos. This ocean of neutrinos even has its own temperature. It is a bit below the temperature of photon radiation, at about 2.2 degrees Kelvin. That, at least, is what theory suggests to us.

EINSTEIN: Now I understand why the mass of the neutrino is of such importance: there are many more neutrinos than nucleons, so that the mass of those neutrinos, even if quite small, can make an important contribution to the mass density in the universe. The same would hold for photons; but photons, we do know, have no rest mass at all.

HALLER: If one of the neutrinos has a mass of just 30 eV in terms of energy units—and that is a bit less than one ten-thousandth of the mass of an electron—we would already saturate the critical mass density in the universe with this contribution. Dark matter would turn out to be neutrino matter; the big galactic clusters would basically be nothing but huge neutrino clouds, where galaxies move about like fish in an aquarium, engulfed in neutrino gravity. Note that the main gravitational field issuing from the cluster would be due to neutrinos.

NEWTON: Suppose dark matter is made up of massive neutrinos. Will they coalesce into those large neutrino clouds we mentioned? Note that would mean a neutrino density in the cluster which exceeds the mean density.

HALLER: You are perfectly correct: inside a cluster, we might easily find as many as 10 million neutrinos per cubic centimeter. Orbiting with a velocity of some 10,000 kilometers per second. And of course, this velocity is below the speed of light.

EINSTEIN: That is an unappetizing thought. Note we are not only

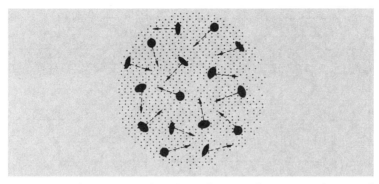

FIGURE 21.5 Schematic view of a galactic cluster. Its dynamics are determined by dark matter—maybe by massive neutrinos or by other massive neutral particles.

in the San Gabrial Mountains, we are also in a galactic cluster. The mere idea that all around us, every cubic centimeter of space is filled with millions of neutrinos that tend to their obscure trade, makes me nervous, to say the least.

HALLER: You are not alone in that. Many experimental physicists who would love to get a clear signal for this ocean of neutrinos feel the same way. But chances are none too hot: the cosmic neutrinos, in contrast to those generated—say, at CERN—have a very small amount of energy. That means it would be difficult for them to trigger any kind of detector. The smaller the neutrino energy, the less likely it is that there is any reaction going on inside. Not surprisingly, it will likely be quite a while before we understand more about the ocean of neutrinos.

To date, we don't even know whether neutrinos have any mass at all. Experimental limits do not exclude a nonzero mass for the muon-neutrino and the tau-neutrino. That might be just large enough to play an important role for the dark matter. It is only in the case of the electron-neutrino that we know its mass to be no bigger than about one electron volt (eV). And that means it cannot be a candidate for the dark matter.

There is another interesting possibility: if the neutrino's mass for all three of its species were, say, 3 to 5 volts, dark matter—or, at least, its neutrino component—could be one third electron-neutrinos, one third muon-neutrinos, and one third tau-neutrinos. Now, should future results from particle physics laboratories tell us that neutrinos do have masses in the range of a few electron volts, this would have an immediate impact on cosmology and, in particular, on the long-term prospects for our universe—and this could serve as an example for the close correlation between the subatomic world of tiniest particles and the macro world of the largest structure imaginable, between particle physics and cosmology. The search for neutrino masses is therefore an important task for basic research in physics.

So, I will summarize our present-day picture of particle physics. There are two sets of particles—those that make up matter in the form of quarks and leptons, and those that make up fields of force, which are responsible for physical forces and interactions: gluons, the W and Z particles, and photons.

From this group of constituents, we should be able to construct a model for fundamental particles and their interactions. That has been done, and the result is the much-touted Standard Model of Particle Physics. That is analogous to Einstein's theory, which is seen as the Standard Model of Gravitational Theory. This theory has been checked with great precision in many different experiments—and never has there been an indication that the theory might be wrong.

EINSTEIN: But still, there is no clarity about the mechanism of mass generation.

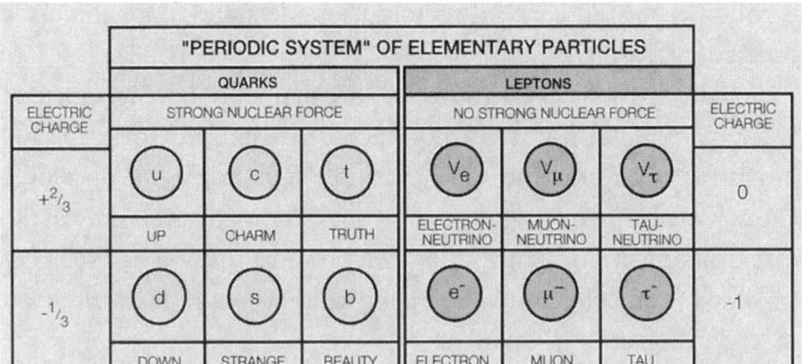

"PERIODIC SYSTEM" OF ELEMENTARY PARTICLES						
	QUARKS			LEPTONS		
ELECTRIC CHARGE	STRONG NUCLEAR FORCE			NO STRONG NUCLEAR FORCE		ELECTRIC CHARGE
$+^2/_3$	u	c	t	V_e	V_μ	V_τ	0
	UP	CHARM	TRUTH	ELECTRON-NEUTRINO	MUON-NEUTRINO	TAU-NEUTRINO	
$-^1/_3$	d	s	b	e^-	μ^-	τ^-	-1
	DOWN	STRANGE	BEAUTY	ELECTRON	MUON	TAU	

FIGURE 21.6 Schematic listing of matter particles. There is a total of three different sets of pairs of leptons and quarks, only the first of which is relevant for stellar matter in the universe. The t and b quarks are alternatively called *truth* and *beauty* (instead of *top* and *bottom*). (Courtesy of CERN)

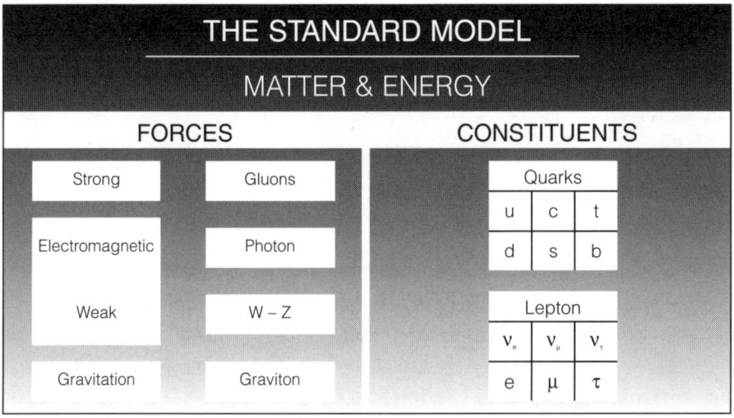

FIGURE 21.7 Schematic arrangement of forces and constituents in the Standard Model of Particle Physics. We distinguish particles of the matter—quarks and leptons—and the particles that carry the forces. See also figure 21.6. (Courtesy of CERN)

HALLER: For now, we just have to live with that. Recall that, just a few days ago, I mentioned in my lecture that we are close to seeing decisive experiments on the matter at the LHC accelerator in Geneva, but not before 2006. For now, it makes sense to use the Standard Model to describe the first moments after the Big Bang.

EINSTEIN: But not here, not now. It will take us a little while to get back to our car, and then there is that dinner at the Athenaeum. That means Newton and I have just a few hours in which to try and digest this provocative food for thought that particle physics offers us, and that you put on our plates this afternoon.

CHAPTER 22

The First Seconds

I would love to know whether God could have made
our world differently—whether logical simplicity
actually would have left that choice.

—Albert Einstein[1]

Next morning, our three interlocutors again got together in the library of the Athenaeum. When Haller entered, he found Einstein already waiting for him; only Newton was still missing.

"Good morning, dear colleague! As we agreed last night at dinner time, this will be the last day of our discussion. The airplane that will take Newton back to Europe leaves before noon tomorrow. I myself have decided to stay in Pasadena for a few days. But still: we really should finish our cosmological discussion by tonight."

HALLER: I don't think that will be too hard. As far as I am concerned, I will fly back to Zurich by way of Chicago.

At that moment their third partner entered the room, and delved right in.

NEWTON: After all that we heard yesterday, the essential dynamics of the cosmos was set in motion—you might say, was preprogrammed—in the first seconds or even microseconds after the Big Bang. That is true at least for its global aspects like the composition of matter, the density distribution, and the like. All that came later on can only be seen as local effects that are of no importance for the global properties of the universe.

HALLER: I agree with you—all the important choices are made in the beginning. This morning, as we take a closer look at the first moments of our universe, I would like us to pay attention to the differences between what we really know and what we extrapolate theoretically.

EINSTEIN: As far as I'm concerned, I go along with that. Still, I'm not opposed to some speculative thought here and there. Our science would not advance without that—even though most of our speculations turn out not to lead anywhere: I'm afraid the wastebasket does remain the most important instrument for the theoretical physicist.

Still, yesterday, you told us that all the interactions of elementary particles can be described in terms of the Standard Model of present-day particle physics—never mind that I find this a terrible name for something that looks like an acceptable theory. I was particularly impressed by the fact that no new type of fundamental interaction has been found in the development of physics since 1930.

In my time, we had the electromagnetic forces, the strong and the weak nuclear ones, and, of course, gravitation; and the same thing is true today. What has changed is how we describe those forces—take, for instance, the description of the strong forces in the framework of quantum chromodynamics. Modern-day particle accelerators put us in the position to describe the structure of matter up to an energy well above 100 GeV. And this means that we can investigate the dynamics of matter down to some 10^{-16} cm, or one thousandth of the diameter of an atomic nucleus. Suppose we apply that knowledge to cosmology. Doesn't that permit us to follow time backwards in the cosmological development until we come close to the Big Bang?

HALLER: We can go back to the very point in time when the temperature of the universe was of the order of 100 GeV. That is in the range of 10^{-8} to 10^{-9} seconds after the beginning, about one billionth of a second after the Big Bang.

EINSTEIN: Excellent. Why don't we just start at that time and try to follow further developments? We can still come back later on and revert to the more speculative part of these discussions, and try to take a peak at what might have occurred in that very billionth of a second.

HALLER: I can, of course, invert the direction of time and follow today's universe backwards to that first billionth of a second.

NEWTON: Go right ahead. What was the world like at that time?

HALLER: In principle, matter was nothing more than a devilishly hot soup consisting of quarks and antiquarks, of electrons and positrons—and let us not forget the other leptons as well as the carrier particles of the forces, the photons and the gluons. The other carrier particles, the W and Z bosons, have a large mass; and that made them play no more than a minimal role at that time, so that they contributed very little to cosmic matter density. That was mostly given by the quarks and antiquarks and by the gluons. And that is why the state of matter at that time is called a quark-gluon plasma.

Today, we are trying to create this quark-gluon plasma artificially, and only for very short times. In particle accelerators such as those at CERN in Geneva, Switzerland, and a new one, RHIC, on Long Island, New York, we use heavy atomic nuclei that we accelerate to high energies before colliding them one against the other. These collisions generate hundreds of particles from the available energy; and this process does have many features similar to what happened at the moment of cosmic creation.

EINSTEIN: Fine, so this is our starting point. The universe expands rapidly, cooling off in the process. But what happened next?

HALLER: After that, what happens is a kind of mass murder—the annihilation of antimatter in the cosmos. Remember, our initial state has lots of antiparticles, in particular, antiquarks. Now, particles and antiparticles annihilate mostly into photons; and of course, they also get created from radiation energy. But as the cosmos cools down, it slowly loses the capability to create antiquarks in the quark-gluon plasma. So they die off relatively quickly, finally leaving a kind of soup of nothing but hot quarks; and these are only the u and d quarks—the heavier ones have decayed in the meantime.

EINSTEIN: And, subsequently, they will try and coalesce into protons and neutrons.

HALLER: That happens at a point in time we can reasonably tell: it happens when the temperature is down to a few hundred MeV, which corresponds to about 10^{13} degrees, or 10,000 billion degrees. And that occurs about one hundred thousandth of a second after the Big Bang. Now the quark plasma transforms into a very hot gas consisting of protons and neutrons: the quarks attract each other with the chromodynamic force. It becomes natural for them to enter into permanent partnership, three of them at a time.

NEWTON: And this process, I presume, starts the formation of structure in the universe.

HALLER: Correct. The synthesis of nucleons from quarks starts the formation of structure in the universe, just as you say; this is the

FIGURE 22.1 Sample collision between two lead nuclei: hundreds of particles are being generated by the collision energy. A precise analysis of such collisions is needed to find properties of the quark-gluon plasma, the state of matter in the early universe about 1 billionth of a second after the Big Bang. (*Photo:* CERN)

first in many steps, which also include the formation of structures as complex as Messrs. Newton and Einstein.

About a second after the Big Bang, the neutrinos get into the act. Immediately after the original explosion, there was a hot neutrino gas present in addition to the quark plasma. Those neutrinos interacted with the particles in their vicinity. They were able to do so simply because matter density was exceedingly high, so that even the weakly interacting neutrinos could not get around colliding with their neighbors—the density of particles in the cosmic soup was still enormous. But one second after the Big Bang, matter density had diminished sufficiently to make neutrino interactions less than a foregone conclusion. So the neutrinos take their leave. From this point on, they inhabit a world of their own—they form a neutrino gas which basically ignores the presence of other massive particles in the universe. The subsequent expansion of the cosmos then results in a successive cooling down of the neutrino gas. By today, our calculations tell us that this gas is now down to the 2.2 degrees Kelvin, as we mentioned before. And that makes the temperature of neutrino radiation a bit lower than that of photon radiation.

The next step of structure formation happens about one minute after the Big Bang. This concerns the synthesis of the light elements, in particular of helium, as we already mentioned. About 300,000 years after the Big Bang, the universe has further cooled down to about 10,000 degrees. At this temperature, the typical energy of a particle is of the order of 1 eV.

EINSTEIN: And that is the temperature which suffices to split atoms into what makes them up—into their nuclei and the electrons of the atomic shells.

HALLER: As time progresses, the temperature decreases, and now we get to the formation of atoms from nuclei and electrons. And that means we have entered into a qualitatively new step of the formation of matter in the universe. From now on, matter no longer exists in terms of electrically charged particles—from now on the building blocks are electrically neutral. And this implies that photons do not interact with the other matter particles except very rarely, because photons "couple" only to electrically charged objects. And so, this stretch of cosmic development is dubbed the time of photon decoupling. Just as the neutrinos did before, this is the time where the photons decouple from atomic matter.

EINSTEIN: From now on, this ocean of photons pervades the universe irrespective of the rest, just like the ocean of neutrinos. Only cosmic expansion acts on it—and that is because the wavelength of electromagnetic waves is increasing, along with all other cosmic distances.

HALLER: The ocean of photons that we observe in intergalactic space today is really a fossil from the early times of our universe—or, more precisely, from a period some 100,000 years after the Big Bang. The humming of that radiation, first perceived by Arno Penzias and Robert Wilson, is the muted echo of the Big Bang. This echo pervades all of our present-day universe. But the energy of the photons in their ocean is steadily decreasing. The radiation that reached our Earth a year ago had been on its cosmic peregrination a year less than what we observe today; as a result, its mean photon energy was larger by about one ten billionth than what we see now. Or what we could see in principle: this effect of cosmic cooling is actually below the level for detection.

The fact that the radiation we see is very homogeneous and isotropic gives us a hint that the energy density in the early universe was very uniformly distributed. Had there been important density variations, we would notice that in the ocean of photons as it appears today.

NEWTON: When we look at our universe today, galactic matter also appears to be distributed somewhat evenly on a grand scale, a scale on the order of a billion light years, which covers about a tenth of the extent of the visible cosmos. But over smaller distances, there is a lot of structure—galactic clusters, the galaxies themselves, and finally, the stars. And that means that in the early history of the cosmos there must have been structure on a larger scale that originated in accidental fluctuations of matter density. But if I understand you correctly, you tell us that such structures cannot have existed just a few hundred thousand years after the Big Bang. And that surprises me. When the ocean of photons decoupled from the matter distribution, I surmise that density fluctuations should already have existed. Normal atomic matter was moving relatively slowly through the universe at that time, and gravity would have generated larger fluctuations in the density of matter distribution. This could have been effected by the gravitational pull of either atomic matter itself or by that of dark matter, such as the hypothetical clouds of massive neutrinos or other particles that we discussed.

HALLER: I agree with you there. Now, our colleagues suspected that there would be low-level fluctuations in the ocean of photons, such as inhomogeneities in the cosmic background radiation. That made them put special detectors in the COBE satellite, capable of noticing even low-level fluctuations. After years of search, such fluctuations were found.

EINSTEIN: Oh really? So we do have something that resembles a snapshot of our universe shortly after it all started?

HALLER: You might call it that. In fact, when the COBE research team reported their finding in 1992, the New York Times qualified it as "God's fingerprint."

NEWTON: What nonsense! These fluctuations are due to nothing more than gravity as it acts everywhere in the universe. There was no need for God's help.

HALLER: Never mind, the COBE snapshot gave us two surprises: for one thing, the fluctuations turned out to be weaker than expected. The temperature of the ocean of photons fluctuates by amounts that are smaller than one ten-thousandth of the cosmic

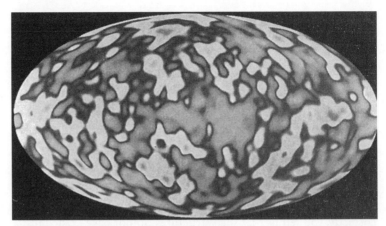

Figure 22.2 Image of the observed temperature fluctuations of cosmic photons. (Courtesy NASA–COBE collaboration)

temperature, which, in itself, is just 2.7 degrees Kelvin. And that indicates variations in matter density at the time when atoms were formed, which were relatively weak. Second, the regions that display the fluctuations are quite enormous by today's scales. Some of them have a diameter of hundreds of millions of light years. And that means that initial fluctuations in matter density also stretched over large distances. To illustrate this: imagine that the universe as we see it today is a sphere with a radius of one meter; on that scale, the first density fluctuations formed over a few centimeters. Further data, in particular obtained by balloon experiments, indicate the presence of density fluctuation at smaller scales. And somehow that led to the formation of galaxies and stars—nobody knows how exactly that happened. Mind my words: that is as far as we understand things today. There are still lots of open questions.

Einstein: This is certainly the way it appears to me. But it is not only the formation of large-scale structure in the universe that eludes our understanding; we can say the same thing about the period from the very start of the Big Bang to the first billionth of a second.

CHAPTER 23

A Cosmic Fairytale

The less a researcher knows, the farther removed he
feels from God. And conversely, the more knowledge he acquires, the closer he feels to the creator.
—Albert Einstein[1]

At noontime, our three physicists were having lunch at the
Athenaeum. Einstein, who really enjoyed the meals at the faculty
club, had ordered the specialty of the house—a large filet mignon,
together with a bottle of cabernet sauvignon. Newton, on the other
hand, quickly finished a salad and got nervous. But Einstein was not
to be bothered. Finally Newton spoke up: "Professor Einstein, there
are a few questions that I simply have to get rid of. Would you agree
with our continuing our discussion right now?"

EINSTEIN: But why not, Sir Isaac—just as long as you permit me
to enjoy the rest of my lunch. I also have a lot of questions—but I
doubt that today's level of scientific knowledge is capable of providing satisfactory answers. And don't forget, we have set a fairy-tale
hour for this afternoon. Haller wants to tell us this scientific fairy
tale, which will contain both truth and fantasy. But if you really cannot wait, fire away!

NEWTON: Right after the Big Bang, after just one billionth of a
second, there was matter in the form of quarks, antiquarks, of leptons and their antiparticles, of gluons and photons, and, God
knows, maybe of other particles that physicists have not found up to
this day. But right afterwards, there began the process which Haller
described as the cosmic mass murder of antimatter, so that, by today,
we are left with practically nothing but matter in the universe.

HALLER: I would even venture to say that the only places in our
galaxy where we can find significant amounts of antimatter are the
CERN laboratories near Geneva and Fermilab outside Chicago.

NEWTON: But we also know that in particle collisions matter and antimatter are produced in equal quantities. So, in microphysics, there is no difference between matter and antimatter.

HALLER: It is a law of Nature—the law of symmetry between matter and antimatter. Antimatter behaves exactly like matter: at CERN, experimentalists made antihydrogen atoms that consist of an antiproton and a positron each. Atomic physics looks the same for atoms and antiatoms.

NEWTON: But this symmetry does give me problems. We should expect that the Big Bang produced all matter from energy by means of pair creation. And that implies that for every particle, an antiparticle appeared. In other words, there should be an equal amount of matter and antimatter.

HALLER: Right you are. If matter-antimatter symmetry prevailed in the universe, the Big Bang would have produced particles and antiparticles in equal amounts. Less than a minute later, essentially all particles would have annihilated upon meeting with their antiparticle; today's universe would be left without matter as we know it—there would be just photons, neutrinos, and antineutrinos. Sounds like a fairly boring universe, doesn't it?

EINSTEIN: In other words, something isn't working here. Symmetry between matter and antimatter doesn't quite work out, unless Nature has devised a clever way to separate them right after the Big Bang; that we cannot exclude. But on the whole, there would have to be as much antimatter as matter in the universe.

HALLER: So it is a fair question to ask whether there are accumulations of antimatter someplace—antistars, entire antigalaxies. But the state of our present knowledge permits us to exclude that idea, at least for the universe as far as we can observe it. Antistars cannot possibly exist in our own galaxy: the annihilation of their antimatter with matter in their vicinity would lead to an incredible production of high-energy radiation. And there is no way that would have escaped detection. But even for distant galaxies, we can say with almost complete certainty that they consist of particles and not of antiparticles. If they, or at least some fraction among them, were made up of antimatter, galactic collisions would lead to a fireworks of radiation. Galactic collisions do occur, and have been observed. But never has one of them been seen to be accompanied by much radiation. We are fairly certain that even the most distant galaxies are made up of matter, that there is no accumulation of antimatter in the observable cosmos. That implies our universe exhibits a drastic violation of particle-antiparticle symmetry.

EINSTEIN: That means our problem moves from macrophysics to microphysics, to the elementary particles—to your specialty, Haller.

HALLER: Until the 1960s, the common assumption among physicists was that symmetry between matter and antimatter is cast in concrete by Nature. But in 1964 a detailed analysis of a short-lived kind of particle called the **K-meson**, which can easily be produced in particle collisions, showed this symmetry not to be fully satisfied. Sometimes it is violated, if only at a small rate, so that it does not show up in most particle interactions. It is an effect of the weak interaction which, you will recall, is mediated by the W bosons. Now, I have to stress that, to date, there is no satisfactory theoretical explanation of this effect.

EINSTEIN: I'm not surprised. The symmetry is ultimately violated, if only a bit; but that may be sufficient to give us the effect we discussed.

HALLER: Still, the apparent violation of cosmic matter-antimatter symmetry and, on the other hand, the almost precise symmetry we observe in microphysiscs between particles and antiparticles, together present quite a problem for our understanding of cosmic evolution. The "software" of our universe—its natural laws—treats matter and antimatter analogously; Nature's hardware, however, doesn't.

To this day, it is not clear what causes this discrepancy between macrophysics and microphysics. There are concrete ideas about it, guessing that a collusion between particle physics and astrophysics can do the job. But to explain this, we have to delve more deeply into the structure of matter. In so doing, we will need to leave the realm of experimentally provable physics.

The small violation of matter-antimatter symmetry that was observed is not sufficient to tell us why there is no significant antimatter in the cosmos. The problem is related to the structure of the atomic nuclei. A nucleon—a proton or a neutron—cannot simply vanish or appear. The number of quarks in our universe is constant—or, more precisely, we know that in every reaction the number of quarks minus that of antiquarks cannot change. This is an important law. It says, for example, that a proton—a bound state of three quarks—cannot decay into, say, a positron plus a few photons. Never mind that such an occurrence would be fatal anyway—all matter would be unstable if it could.

NEWTON: Frankly, I had never thought of that kind of proton decay. But you're right: it is odd that the proton is so stable. Are we sure that this decay is really excluded?

HALLER: It has been looked for, but in vain. The experimentalists tell us that we can guarantee proton stability to a lifetime of 10^{32} years. And that means, I don't have to tell you, their lifetime exceeds that of our universe by 22 orders of magnitude.

NEWTON: How on Earth can one make such a statement? Matter has been around for some 10 billion years. That should preclude us from saying a proton lives so much longer.

EINSTEIN: But we can, Sir Isaac, quantum theory permits it. It is just like the decay of a neutron: this particle lives on average for ten minutes. However, we are really talking averages that are as valid for all neutrons as when an insurance company says that a man lives on average for seventy-five years. There are neutrons that decay just a few seconds after being released by, say, a reactor. Tough luck for these, you might say. In the same way, we can look for proton decay by trying to follow the fate of as many as possible so that the poor specimen that has "tough luck" will be seen. You may say this sounds like a game of roulette; but all of quantum theory is really nothing but a gigantic roulette game—and I will admit that it is not my favorite theory.

HALLER: This is indeed the way the lifetime is determined. To speak of a lifetime of 10^{32} years, at least 10^{32} protons have to be looked at—one for each year. To do so, we need a few thousand tons of matter; ten thousand tons of water will do, and we'll get back to this.

Anyway, proton stability is guaranteed in the Standard Model of Particle Physics. Thus, the number of quarks in today's universe must be the same as the number of quarks minus that of antiquarks right after the Big Bang.

EINSTEIN: Those are strong words: in the beginning, the universe was full of antiquarks and quarks. And how should the universe know that the difference between those two quantities should be exactly what we see today, after the demise of the antiparticle? I cannot see that as a viable solution.

HALLER: You are not alone there. But maybe there is another way: maybe the number of quarks minus antiquarks changes after all, as the cosmos evolves. In that case, the difference could have started out by being zero, as we would naively expect. This, however, would need processes that go beyond what the Standard Model of Particle Physics permits.

Over the last twenty years, particle physicists have theorized that quarks, electrons, neutrinos—all of them apparently structureless constituents of matter—are nothing but different incarnations of one and the same "ur"-particle. Differently put, it says that the quarks, electrons, and neutrinos are closely related.

Now, we have several reasons to believe that. One of them is based on the structure of electric charges. Let's look at the constituents of normal matter in the universe: we find the u and d quarks—a proton consists of two u-quarks and one d-quark. The u-quark has an electric charge that equals two-thirds of the electric charge of the positron which, you may recall, we call the elementary electric charge. The charge of the d-quark is negative and equals one-third that of the electron. When we look at the two quarks u and d, and then at the two leptons (the electron and its neutrino), we notice that the sum of all the electric charges add up to zero if we also include in our addition that there are three colors of quarks. The sum of charges then looks like this: 3 times $(2/3 - 1/3) - 1 = 0$. Now, we consider those two quarks, u and d, plus the electron and its neutrino as making up a lepton-quark "family." And we know that the same picture recurs twice more with heavier quarks and leptons. There are all together three quark-lepton families.

NEWTON: It looks like the number three has a special standing here. The quarks have three colors, and the lepton-quark family appears in three versions. As far as I can tell, the number of colors and the number of families are not at all related.

HALLER: Right—this is one of the great mysteries of today's microphysics. Whence the number three? Nobody knows.

EINSTEIN: Don't sound so pessimistic, Haller! You should be pleased that the number three has this refreshing surge in popularity in our cosmos. Three colors, three families, and all that in three spatial dimensions—that makes up three mysteries, doesn't it? Two of them sound too modest; four—let's not exaggerate. Three sounds just fine.

HALLER: Now please tell me: why should the number of families equal that of colors, and why should that again be the same as the number of space dimensions? Just by coincidence?

EINSTEIN: I would not know. But I'm sure we will find out. If I were in your place, I would be pleased about every unsolved problem—it gives today's particle physicists something to explain. Now, please, continue your story.

HALLER: All right. For starters, let's stick with the first family. To explain why the sum of electric charges of all the members of our family adds up to zero, we assume that leptons and quarks are related to each other by means of a basic symmetry principle. The fundamental forces among the elementary particles are described in terms of three basic interactions—the electromagnetic plus the weak and strong nuclear interactions. We'll leave gravitation out of this consideration because it is too weak to be noticeable between individual elementary particles.

The strengths of the elementary forces of Nature are quite different—the strong force is ten times as strong as the electromagnetic force. Still, we can imagine that all three basic forces are different manifestations of one and the same elementary force. We see a unification of the forces of Nature, which manifests itself only at very high energies, such as right after the Big Bang.

Theories that incorporate this view do not stop at unification of the forces. They also predict that the difference between the various basic particles—between electrons, quarks, and their antiparticles—will vanish. In other words, electrons, quarks, and their antiparticles are reduced to the status of being different manifestations of one and the same building block of matter. At that level, we are left with one matter particle which, like a chameleon, pops up as a quark, as an electron, or as a neutrino—as the case may be.

In the first family, there are six different colored quarks, a total of eight particles. Then there are the eight antiparticles to these. So we have sixteen different objects altogether.

EINSTEIN: Now here is another interesting number: you get sixteen when you multiply the number 2 four times: $16 = 2^4$. Maybe this gives us a key to this problem?

HALLER: At least you are on the right track. If you ask yourself which symmetry is able to establish family relations among sixteen different objects, a mathematician has no trouble giving you an answer: it is the symmetry that you get when you consider all rotations in a ten-dimensional space.

NEWTON: Ten dimensions? I will admit that just imagining rotations in normal three-dimensional space gives me plenty of trouble.

HALLER: There is no need to really imagine that. This ten-dimensional space is just a mathematical construct that is capable of establishing a relation between those sixteen elementary particles and the elementary forces of Nature. It really is an abstract symmetry, so to speak.

EINSTEIN: Indeed, what you say sounds quite abstract. Still, it will be interesting to see that this symmetry of yours puts quarks, antiquarks, electrons, neutrinos, and their antiparticles on one and the same level—establishing, if you will, a basic subnuclear democracy.

HALLER: I like to mention that this democracy brings in a problem: the symmetry we are discussing includes the positron. This particle has the same electric charge as the proton, as required by the symmetry. A grand unification of the fundamental forces—should it be truly realized in Nature—would certainly explain the quantization of charges: it enforces the rule that the sum of electric charges adds up to zero in each lepton-quark family. We already know that this is the case.

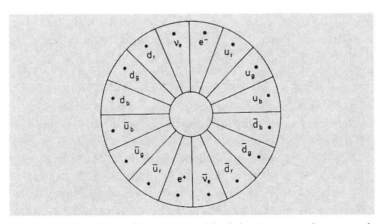

FIGURE 23.1 The sixteen elementary particles belonging to one lepton-quark family that are related in higher symmetry schemes. The relevant symmetry operation is a rotation in a ten-dimensional space.

But here comes the problem: if positrons and quarks are really closely related, there is no reason why the proton should be stable. We would expect it to decay immediately into a positron and one, or maybe several, photons.

EINSTEIN: This topic gets more hairy when we look at a hydrogen atom—at a proton with one electron in its orbit. If the proton decayed, with the positron carrying away its charge, we would imagine there is a fair chance that this positron will collide with the orbiting electron, so that the two annihilate into radiation. The hydrogen atom would vanish into radiation and nothing else.

NEWTON: Now, hold on! Our universe is comprised largely of hydrogen. If the proton were unstable, hydrogen could vanish into radiation. Does that not imply that, conversely, the quarks and electrons that make up hydrogen could have been generated by pure radiation? Every process in Nature is, in principle, reversible. The inverse of decay is creation. And that would give us the key to understanding the creation of matter from pure energy.

HALLER: Sir Isaac, I have to compliment you; you just took the words out of my mouth. Proton decay or, more generally, the decay of matter needs to exist if we want to understand the creation of matter from the energy of the Big Bang.

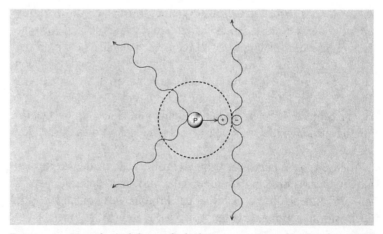

FIGURE 23.2 Hypothetical decay of a hydrogen atom into photons: the proton decays into a positron and two photons. The positron annihilates when hitting the electron of the atomic shell, generating two more photons. The entire hydrogen atom has thus been turned into radiation.

EINSTEIN: Just like life: birth and death are, in a way, two sides of the same coin. One cannot be without the other. You were talking about a problem before—but as we just saw, the decay of matter solves the problem of creation. So where is the problem?

HALLER: Maybe I should not have called it a problem. We don't observe proton decay. That means the grand symmetry that ties quarks and leptons together is strongly broken. The implication is that the unification of fundamental forces will become a reality only at very high energies, if at all.

EINSTEIN: Hearing you say "at very high energies," I surmise that these energies must be gigantic.

HALLER: And so they are—they amount to something like 10^{16} GeV. This energy corresponds to the rest energy or mass energy of 10^{16} protons, 10 million billion protons.

EINSTEIN: And that is the mass of a germ. You're not being modest, Haller, this energy is certainly impressive. Recall that we were talking about length scale corresponding to Newton's gravitational constant—in other words, Planck's elementary length once we bring quantum theory into the game. What we did not mention at the time was the fact that this length scale is tied to a corresponding energy scale. And this correspondence gives us a value of 1.2 times 10^{19} GeV. I recall Planck mentioning to me that he had not the foggiest notion of what to do with this incredibly high energy value. And now you are telling me that the grand unification of leptons and quarks happens somewhere around 10^{16} GeV—just a factor of one thousand below the number that made Planck shiver.

HALLER: It is quite possible that there is a connection—a bridge, if you wish—between Newton's macroscopic constant and particle physics. I can tell you that today's physicists, with the view on this possible connection, are not at all unhappy when they see the energy scale where the forces of Nature become equally strong turn out to be so enormous—never mind that this fact minimizes chances for an experimental test.

EINSTEIN: The CERN accelerator barely manages to reach 10 millionths of a billionth of this energy. In other words, if we believe in this grand unification scheme, we have to extrapolate over 14 orders of magnitude. That is comparable to an extrapolation from the size of an apple to the size of an atomic nucleus—and you'll agree that this indicates an amazing speculation. Do you really believe that this enormous step makes sense? Fourteen orders of magnitude—anything can happen there.

HALLER: It can, yes; but it needn't. Think of the atom: the diameter of an atom is 5 orders of magnitude larger than the diameter of the nucleus, and nothing much happens. So why couldn't that be true also for 14 orders of magnitude.

NEWTON: I consider that highly speculative; but then, why not? We might try. Haller does predict that the proton is not stable and that, in principle, we might see its decay.

HALLER: The unification of the forces of Nature at high energies does not happen automatically; it is guaranteed by a new, fourth force which, you might say, acts as a mediator among the other forces. This one is often called the X-interaction, and its carrier particles X-particles. The main difference between the X-particles and the other force carriers, such as photons, is their mass. We know that photons are massless, but X-particles are endowed with an enormous mass, the 10^{-8} grams we mentioned above. And still, notwithstanding such huge differences in mass, the unified theories treat the carrier particles of all the forces as a family, and that includes photons as well as X-particles.

EINSTEIN: An odd family it is, with the midget massless photon and the gigantically heavy X-particle! Still, this huge mass difference also gives us a measure for the breaking of the symmetry; it tells us something about the looseness of the family ties between quarks and leptons.

HALLER: That is one way to put it: the larger the mass of the X, the more strongly the symmetry is broken. This mass ultimately also determines the lifetime of protons, because the X-interaction has a peculiar property: it is capable of changing an electron into a quark, even into an antiquark. That means that at high energies, as soon as this X-interaction comes into the game, the differences between electrons and quarks will vanish. And this gives us a chance to understand how, shortly after the Big Bang, electrons and quarks were generated—but not particles and antiparticles in equal amounts.

True, right after the original explosion, there were as many electrons as quarks, and the antiparticles of both. But then an asymmetry that was initially almost unnoticeable tipped the balance between matter and antimatter by a tiny amount in favor of matter:

for about a billion particle-antiparticle pairs, there was one additional particle. Shortly thereafter, there was the mass annihilation among particles and antiparticles. Only those few particles that did not find a partner to annihilate with survived this cosmic inferno. And these are the electrons and quarks that make up today's matter, including our bodies, my friends.

EINSTEIN: I understand—a tiny effect becomes gigantic, simply because of the annihilation of all the particle-antiparticle pairs. It ensured the dominance of matter in our universe as it is.

HALLER: To ensure this mechanism, we need two things: first, the tiny violation of particle-antiparticle symmetry of which we said that it was first seen in the case of K-mesons, and second, the family relationship between leptons and quarks which, however, also implies the possibility of proton decay. This means that the same mechanism which today permits us to interpret the formation of matter during cosmic evolution also predicts catastrophic consequences for the future. It implies the disappearance of matter into radiation.

Unfortunately, we will never be able to study in our labs whether there is, in fact, a grand unification of the fundamental forces at high energies. An accelerator capable of giving the energies needed for the unification to individual particles would have the dimensions of our galaxy. Only at the time of the Big Bang can there have been particles of the required energy.

Things look more promising for proton decay. Our Japanese colleagues are about to increase the size of the Kamiokande detector such that one can observe 100,000 tons of water over a number of years. If protons decay with a lifetime in the range predicted by the unification model, that will most probably be seen. And the observation will set an important milestone in the history of the sciences.

EINSTEIN: Let me try to summarize: you tell us that, today, we have an approximate picture of cosmic developments right after the Big Bang: at first, the cosmos was made up of a very hot elementary-particle plasma, at enormously high energies. For the briefest of moments, it existed in the state of highest symmetry. There was no difference between quarks and leptons, nor between the different forces of Nature—with gravitation being the only exception.

The cosmos then expanded and cooled down. The high degree of symmetry was broken. Differences developed between quarks and leptons, and between the different interactions. Particle-antiparticle symmetry was violated by a tiny effect, causing about one additional quark to appear for every billion quark-antiquark pairs. Fractions of a second after the Big Bang, those quark-antiquark pairs annihilated into radiation. The unpaired quarks could not do that—so they remained in the universe and became the building blocks of emerging galactic matter.

NEWTON: The picture you are drawing has another inescapable consequence: today's matter will vanish with time: in some very distant future, the universe will be nothing but an ocean of photons and neutrinos. There will be no planets, stars, galaxies. Nothing will be left to bear witness to the diversity of matter distribution in the cosmos we see today, some 15 billion years after the Big Bang.

EINSTEIN: We should not forget about the black holes. They survive the decay of matter, but they are nothing but the tombstones of matter past in spacetime—and they, too, will ultimately vanish. These are not rosy prospects for the future! Agreed, that is not a future we have to worry about on a human scale.

Still, let us not forget that the picture you paint, Haller, contains a lot of speculation that is not supported by experimental tests. You tell us a cosmic fairy tale which might well bear little resemblance to reality.

And look: we are the only ones left in this restaurant. Time to get back to the library! If it's all right with you, let's meet there again an hour from now. I first need to get rid of the hum in my head that all these speculations have set into motion.

Exactly an hour later, the last round of these discussions started in the Athenaeum library.

EINSTEIN: I well recall the question I put to myself after our first conversation, here at Caltech with Hubble, concerning the cosmic expansion he had discovered: why is the universe expanding altogether? What is it that drives space apart; what is that cosmic yeast?

FIGURE 23.3 (*opposite page*) An image of cosmic evolution in time—from simplicity to complexity, from quarks and leptons to living beings to humankind. Matter in the early universe consisted of a very hot gas of elementary particles without structure. Subsequent development formed nucleons, nuclei, atoms.

The Big Bang

15 thousand million years

1 thousand million years

300 thousand years

3 minutes

1 second

10⁻⁴ seconds

10⁻³⁶ seconds

10⁻⁴³ seconds

3 degrees K

18 degrees

6000 degrees

10⁹ degrees

10¹⁰ degrees

10¹⁶ degrees

10²⁷ degrees

radiation
particles

heavy particles
feeling the weak force

quark
anti-quark
electron

positron (anti-electron)
proton
neutron
hydrogen
deuterium
helium
lithium

Modest observers of cosmic happenings that we are, we can do no better than proceed in the opposite direction. We observe cosmic expansion; and we extrapolate backwards in time, to the Big Bang. Up until now, I have refrained from asking the question, afraid as I am that we still have no answer. But now that we are coming to the termination of our discussions, I have to ask: Why, Haller? Why was there a Big Bang in the first place? Was this really the beginning of the universe, or was it ultimately nothing more than another episode in cosmic history, if a rather violent one?

NEWTON: The natural sciences are not able to explain everything. It is God who sets the rules of Nature. It is He who holds the keys to our universe. He had no trouble setting off the Big Bang. After all I heard, this explanation is good enough for me.

HALLER: But it shouldn't be, Sir Isaac. There are a few more ponderous problems that cosmology poses us. Let me just mention two of them, both of which carry a name: the first one is called *the problem of flatness*. We know that there is a critical density of matter in the universe. If the actual density is smaller than this critical amount, the universe will expand forever. If it is larger, the Big Bang will happen in reverse sometime in the distant future. The universe will collapse.

EINSTEIN: We already mentioned that matter density in our cosmos is somewhat below the critical value—but not by much. As a consequence, the curvature of space in the universe is quite small. It may even be vanishingly small.

HALLER: And here is the problem: why are actual matter density and critical matter density so close to each other, within a factor of 10 or so? They have nothing to do with each other. They could differ by many orders of magnitude.

NEWTON: Could it not be a simple coincidence?

HALLER: I cannot exclude that. But that would be an extraordinary piece of good luck. Whoever set the cosmic machine in motion must have programmed the Big Bang and the tiny fractions of seconds that followed it in the most precise manner, so that our universe developed in the way we are observing today. I cannot believe that this is a matter of coincidence. I might do so if there were not our second problem. This one is called *the problem of the horizon*; it deals with the uniformity, or smoothness, of the cosmic background radiation.

EINSTEIN: When we recently talked about that, it made me think that the uniformity of this radiation is really mysterious. Let us look, for instance, at radiation that comes from the direction of Ursus Major. At the same time, somebody in Cape Town observes the radiation coming at him from the direction of the Southern Cross. In both cases, we measure a temperature of about 2.7 degrees Kelvin. And in both cases, that radiation has been on its merry way for about 10 billion years. But the regions of the universe from which it comes at us have never had anything to do with each other—not even right after the Big Bang.

Now, there is no way to understand why that radiation is so uniform. We would expect that, in as catastrophic an explosion as the Big Bang, different regions of the emerging but rapidly expanding universe would develop in different ways. But no such thing is seen. It looks as though the Big Bang was just a gentle act of creation, with the creator seeing to it that no asymmetries are able to develop.

HALLER: And that is exactly our problem. The electromagnetic waves that come at us from different parts of the universe never had a causal interaction. They have, as we say, different horizons—and mind you, this has nothing to do with the horizons of the black holes that we have discussed. Therefore, we see no reason for the complete isotropy and homogeneity, just as we don't see any reason why the average matter density is so close to the critical density.

NEWTON: Maybe those two phenomena are somehow related?

HALLER: That is entirely possible if we accept the explanation I will now propose. As I promised before, I will come back to Einstein's cosmological constant on this occasion.

EINSTEIN: The one I gave up on, since it didn't look useful to me.

HALLER: But that looks somewhat more promising now, as I will show. I want to get back to the problem of mass: we mentioned that the masses of all particles are possibly due to their interaction with a field that is all-pervasive, that we call the Higgs field. That includes the masses of the hypothetical X-particles, which we mentioned in connection with proton decay.

EINSTEIN: But what on Earth is the connection between my cosmological constant and the Higgs field?

HALLER: A lot, if I am right. The masses of the particles are generated by the breakdown of a symmetry. This breakdown is a bit like what happens when water freezes: when water solidifies into ice, it

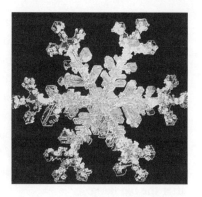

FIGURE 23.4 The hexagonal structure of snow crystals is the result of broken symmetry. Other than in isotropic water, these crystals define certain preferential directions in space.

gives up heat—indeed the same amount of heat as would be needed to melt that ice back into water. Water is a very homogeneous substance; it has a high degree of symmetry. But when ice crystals start forming they destroy this symmetry, because these crystals have a certain preferred direction in space. Just recall the hexagonal structure of snow crystals.

NEWTON: That means there is a connection between symmetry-breaking and the release of energy, as in the mechanical model for the Higgs phenomenon that you mentioned in your Berlin lecture. The sphere rolls down from the tip of the Mexican Hat, down into its rim. Energy is being freed, but the symmetry is broken.

HALLER: But now to Einstein's cosmological constant: let us suppose that, in the beginning of the development of our universe, there is, all of a sudden, the Higgs field in its unstable but highly symmetric state. All masses are still zero. In this state, the vacuum, or if you wish, "empty" space, has a great energy density. Now the universe is expanding. In contrast to normal energy density, the energy density of the vacuum does not go down as the space expands. That means the rate of expansion remains the same as the universe grows in size. The larger a given volume in the universe, the bigger the amount of energy that propels the universe's expansion. The cosmos behaves like a predatory animal whose hunger increases as it devours its prey. That is what leads to its demise: the animal will perish from overeating.

EINSTEIN: My cosmological constant seems to have similar properties. It propels the expansion of the universe just like the Higgs field in its unstable state.

HALLER: And that is exactly what I am driving at. As long as the Higgs field is in its unstable state, we have a universe with a large cosmological constant. In other words: cosmology has reinstated your term—but in a way that you could not possibly have foreseen.

As long as this cosmological constant keeps driving the expansion of the universe, the result is striking: the universe expands at an amazing rate. This rate is so tremendous that physicists no longer call it expansion, they now call it *inflation*. Due to the power of symmetry-breaking, which manifests itself by the huge differences between the masses of different particles, there is a doubling of the cosmic volume every 10^{-35} seconds immediately after the Big Bang, and that happens about a thousand times, up to age 10^{-32} seconds of the universe. In this short time period the cosmos has expanded by a factor of 10^{50}. This means that the entire universe we see today is descended from a tiny region of space which, when inflation began, had a diameter of just 10^{-35} light seconds.

EINSTEIN: I see where you are leading: it is the inflation trick that solves our problem with the horizon. Inflation sees to it that all regions of the visible universe were actually connected when this process started. It drove space apart and, in the process, made sure the universe would stay homogeneous. And now we have a reason for the homogeneity of the cosmic background radiation.

HALLER: And we also have an automatic solution for the problem of flatness. No matter how strong the curvature of the universe when

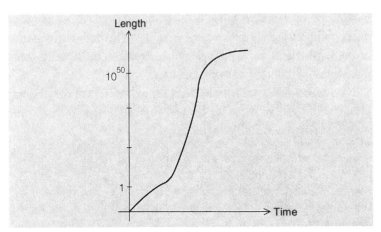

FIGURE 23.5 A given length changes by many orders of magnitude in the model we call the inflationary universe.

inflation began—the strength of inflation saw to it that the curvature is no longer noticeable. Think of a soccer ball: you see the curvature of its surface immediately. But if we managed to inflate it by a factor of 10^{50}, we would not be able to notice any curvature in its surface: for all intents and purposes, its surface would look like a plane.

EINSTEIN: That means we would expect our universe to look flat without curvature.

HALLER: That would be the consequence. There is no space curvature, and the density of matter equals the critical value, just as Newton wanted it.

EINSTEIN: Not bad at all! We have already seen that matter density is possibly close to its critical value, if only we include dark matter in these considerations. But the idea of inflation does give me a problem: once it gets started, it would be hard to stop. Now, you told us that inflation did cease some 10^{-32} seconds after the Big Bang. What would make it stop?

HALLER: During inflation, the cosmos is in an unstable state; this situation cannot go on forever: Nature always wants to approach the state of minimal energy, and that is the state of broken symmetry. Recall the Mexican Hat. In the various models that have been developed to describe such a process, this would happen about 10^{-35} seconds after the Big Bang. Once the state of broken symmetry has established itself, inflation stops. During the transition to that new state, a great energy density is being freed—it is, we might say, the heat freed by the melting of the vacuum. This energy density is manifested by a very hot gas of elementary particles, of leptons, quarks, photons, gluons, W and Z particles. The high temperature of the Big Bang is now being generated by the transition into this new state—before that, the universe was cold.

NEWTON: So now we have gotten to the point where the original plasma soup of particles was cooked—the soup that we had presupposed when we originally discussed the Big Bang.

HALLER: Exactly. The transition from one vacuum state into another gives us the tremendous power of the Big Bang which, failing that, we would have to declare an act of God. But I have to stress that many details of the mechanism that causes inflation to stop are still unclear.

There is an interesting speculation that inflation was the result of quantum fluctuations of a field that need not have been identical to the Higgs field. In this speculation, the universe reaches out to infinity. In a few regions, spontaneous fluctuations of the field cause a large cosmological constant to appear; this constant leads to strong inflation in that region, generating a bubble that keeps inflating.

That notion would make the universe look like an infinitely large system of boiling water in which bubbles keep forming. Our universe that we can look at through a telescope is only one such bubble—it is thus only a tiny part of a much larger cosmos. But like all bubbles it does not remain forever. Somewhere in the distant future—after a stretch of time that is enormous even in comparison to the present age of our universe- the bubble will collapse.

NEWTON: In other words, we may have to start distinguishing between the cosmos and the universe—e.g., our universe. The cosmos will exist for all eternity, and its extent goes out to infinity; the universe exists for a very long but limited time: the Big Bang itself is nothing but the spontaneous formation of a bubble universe.

HALLER: Looking at cosmology in this context, we can also imagine that it was quantum fluctuations that led to macroscopic phenomena after inflation. In this fashion, we could establish a link between quantum phenomena and the gigantic fluctuations of density that we see in much of the universe, like the formation of galactic clusters or the low-level fluctuations in the cosmic background radiation that were observed by the COBE project.

NEWTON: So far, we have not mentioned how the elementary Planck units of space and time are to be introduced in cosmology. His unit of time, which amounts to 10^{-43} seconds, is fixed by the gravitational constant and the constant of quantum theory. What happens when the universe passes that age of 10^{-43} seconds?

HALLER: As you know, we have no trustworthy theory that combines quantum phenomena and gravitation. Maybe it simply doesn't make sense to talk about time before this Planck era: space and time were subject to quantum fluctuations before that.

Imagine space and time as though they made up the surface of the ocean. From some great height—say, from the window of a plane during a transatlantic flight—the surface of the ocean looks completely level, completely even. When the plane loses altitude, we first notice wave formation, and as we keep approaching, we distinguish the actual structure of the waves and the foam on their crests. It makes sense to assume that space and time might also have a com-

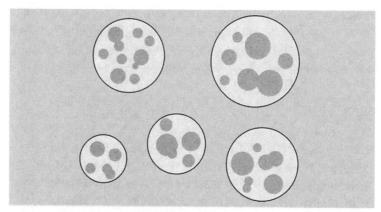

FIGURE 23.6 A hypothetical cosmos that consists of a number of expanding bubbles, each of which forms its own universe containing galactic clusters.

plicated structure once we get down to the order of magnitude of the Planck units. Some physicists will even tell you about "spacetime foam," which is caused by quantum physics.

Even notions such as "before" and "after," which have a clear meaning when we talk about time, now in this instance become questionable. Many physicists tend to believe it makes no sense to even speak of a time before the Planck time. The history of the universe starts only as space and time have overcome their turbulent youth and have reached the stage, so to speak, of adulthood.

After a short pause, Einstein said: "Gentlemen, I think we have now come to the limits of what research has established; and that brings us to the end of our cosmic fairy tale. The future will tell how much of it was reality, how much fantasy. But there is one thing I would still like to stress before we close our discussion: cosmic development ever since the Big Bang exhibits a fascinating interplay between space, time, and matter. Whatever we find in our universe today is the frozen remainder of a very hot phase right after the Big Bang, when all matter formed a very hot gas of elementary particles without the slightest trace of either permanence or structure, accompanied by a steady sequence of production and annihilation processes.

"When this plasma cooled down, structure emerged. The many-faceted and colorful appearance of the world as we know it is due to this interplay of symmetry and symmetry-breaking, but also of the very precise way in which the laws of Nature collude—and, to date, we honestly don't understand many of the details of this game.

"But let me also direct your attention to the harmony between these laws of Nature: it took a highly ordered and precisely tuned interplay that made them come up with the universe of which we are a part. It gives us the impression that the universe knew from the start that, at some time, there would be creatures such as you and I, capable of tracing the secrets of its beginnings. Somebody once asked me whether I believe in a creator of this universe. I can give you my answer without quoting myself verbatim: I do believe that a religious attitude is the same thing as the recognition of harmony in the existence of all things—a recognition that there are impenetrable truths that contain an ultimate reality of radiant beauty. We may catch no more than dull glimpses of it—but knowing that, and foreseeing those glimpses, that is what makes up the basis of a true religious attitude."

Next morning, our three colleagues enjoyed a lengthy breakfast in the Athenaeum. They were to part company. Haller and Newton planned to leave Pasadena the same day. When Haller directed his rental car from the parking lot into Hill Street, there was Einstein standing in the archway of the Athenaeum, waving with both hands. Haller returned the salute as best he could without losing control of his car. He was determined never to forget this last image of the great scientist.

CHAPTER 24

Epilogue

"Get up, Professor! You were supposed to be on the road by nine o'clock!"

Adrian Haller heard this admonishment from the housekeeper who was standing at the bottom of the stairs. She had awakened him from an uncommonly deep sleep. He looked at his alarm clock and was horrified to find it showing 9:30 A.M. He woke up fully, only to notice that he was not in the Caltech Athenaeum, but in Einstein's house in Caputh. That is when the memory of his barely completed dream caught up with him. A short time later, as he sat at the breakfast table by Einstein's fireplace, he cast a few furtive glances about the room—but there wasn't the slightest trace of either Einstein or Newton.

When the taxi driver rang the doorbell, it was half past ten. Still, it did not take Haller long to get on his way. The car took the road through the forest before driving by the shores of Lake Templin in the direction of Potsdam. In the backseat, Haller was mentally running through the talk he was going to give at Potsdam University—a new institution that dated back no further than German reunification in 1990. The title of his lecture was to be: "Einstein's Ideas on Gravitation and Elementary Particles."

APPENDIX

The Basic Tenets of the Special Theory of Relativity

In my book *An Equation That Changed the World*, I presented Einstein's Special Theory of Relativity—in particular, the most important consequence of this novel interpretation of space and time: the equivalence of matter and energy. That theory is the foundation on which Einstein later constructed the General Theory of Relativity, which I introduce in the present book. Many of the aspects of the previous topic are not important for the phenomenon of gravitation. The reader therefore does not need a detailed knowledge of the special theory in order to find his way in the building of the general one. This means he needn't have read my previous book in order to understand the present one. But I found it useful to present in this appendix the basic tenets of the Special Theory of Relativity.

This theory was completed by Einstein in 1905; it revolutionized the interpretation of what we mean when we speak of space and time. Newton's classical theory tells us these concepts are universal, and they do not depend on the state of motion of the observer. The flow of time is universal anywhere in the cosmos. Newton stressed this by speaking of "absolute space" and "absolute time."

When, toward the end of the nineteenth century, it was found that the velocity of light does not depend on the specific state of the observer—as we would have expected in the framework of Newton's mechanics—Einstein was the first to notice that this odd phenomenon can be understood only if we interpret both space and time in a new way: the velocity of light in a vacuum is a universal constant of Nature and has the same value irrespective of the reference system.

This phenomenon contradicts Newton's mechanics. It implies that space and time are not independent of the reference system: if they were—and we recall that Newton took this for granted—we would have to conclude that there cannot possibly be a universal velocity of light.

In Einstein's new understanding, the flow of time is different in a reference system that is at rest and in another one that is in motion relative to the observer (who may, for instance, be sitting in a moving train). Similarly, the structure of space is dependent on the observer: the length of an object, similarly to the flow of time, depends on the state of motion. This concept differs from classical mechanics concerning space and time only then in a noticeable way when the velocities in the system are not all too small when compared with the velocity of light (which is about 300,000 kilometers per second, or 186,000 miles per second). For this reason, the effects of Einstein's reinterpretation of mechanics can be neglected when we consider a moving car; but we *must* heed them when talking about a rapidly moving particle in an accelerator—say, for example, in the Tevatron at the Fermi National Accelerator Laboratory outside Chicago. In these accelerators, the velocities of the particles are not much slower than the velocity of light (in fact, they are within less than 1 percent of that maximum velocity).

One of the consequences we have to adopt from the Special Theory of Relativity is this: no material object in the universe can move at a velocity faster than that of light. This consequence also contradicts Newton's theory, which admits no such thing as a maximum velocity for the motion of material objects. Another very important consequence of Einstein's theory is that the mass of a material object is simultaneously a measure for the energy content of that object. In Newton's mechanics, on the other hand, the mass of, say, a particle is a universal and indestructible property whereas, in Einstein's theory, it is just one special form of energy. Under the proper conditions, mass can be changed into energy—for instance, into the radiation of light. In such transformations, Einstein's famous equation $E = mc^2$ constitutes the rate of exchange between mass and energy. To give an example: a proton and its antiparticle, the antiproton, form a system with a mass of 3.346 times 10^{-27} kilograms. When we bring these two particles into contact, they will annihilate into energy, generating in the process photons, the particles of light. The sum of the energies of the photons equals, according to Einstein's prescription, 1,877 million electron volts (or, for short, MeV). In particle physics, it has become customary not to use units of mass, like kilograms, for particles—rather, particle masses are quoted directly in terms of energy, taking Einstein's relation for granted. The mass of the electron is simply quoted as 0.511 MeV, that of a proton as 938.3 MeV.

Einstein's new interpretation of space and time can be seen like this: we describe our three-dimensional space in terms of a coordinate system that has three axes called the x, y, and the z axis, at right angles to each other. Every point in space is uniquely described by its three coordinates. The square of a straight line joining points A and B is given by

$$\ell^2 = (X_A - X_B)^2 + (Y_A - Y_B)^2 + (Z_A - Z_B)^2,$$

where X_A is the x coordinate of point A and so on. The length ℓ is given in terms of the coordinates, but depends only on its two endpoints, irrespective of the coordinate system chosen. We may move that system in any direction, we may turn it around at any angle—the length will remain the same. It is a quantity independent of that system. This fact is often described by saying: *the spatial distance between two points is invariant with respect to changes of the coordinate system.*

As we accept the tenets of the Special Theory of Relativity, this feature changes when we have moving reference systems, or moving coordinate systems. The distance between two points in space is no longer an invariant quantity. It depends on the velocity of the observer in motion. For an observer in motion, the distance he sees appears shorter. The reason for this astonishing phenomenon is to be found in the fact that the Special Theory of Relativity includes time in the discussion. Three-dimensional space must now add time as an added dimension. Space and time are no longer independent quantities as in Newton's mechanics—they are comprised in what we correctly call *spacetime*. The points in this spacetime continuum—and by that we simply mean points in space at a given time—are called *events*. An event is characterized by a point in space and by the time at which it happens. Think of the form you fill out when entering a country; it asks for an "event," the date and place of your birth.

An important implication is that in spacetime—out of a totality of all possible events—we can fix a distance between two events, just as we defined the distance between two points in three-dimensional space. Two events are uniquely characterized by four coordinates each. The square of the distance between the two is then given by

$$s^2 = c^2(t_1 - t_2)^2 - (x_1 - x_2)^2 - (y_1 - y_2)^2 - (z_1 - z_2)^2$$

(where, of course, c is the speed of light).

As you can easily see, this so-called relativistic distance between two events can be either positive or negative, depending on the relative size of the different contribution of space components and time component. It will be positive when the time component prevails. To give an example: the distance between two events that happen in the same place—take, for instance, the distance between the Munich Oktoberfest in 1995 and that in 1996—is positive. We call the distance between these two events timelike. If the distance between two events turns out negative, the space components will prevail, and we speak of events that are at a spacelike distance with respect to each other. The New Year's Eve celebration in Munich on December 31, 1995, was at a spacelike distance from the analogous event in Hamburg.

We have one important feature: *the distance between two events that are joined by a light signal is zero.* In this case, the spatial part and the time part of the distance cancel each other. The distance between the event "send light signal from Mount Palomar Observatory to the Moon" and the event called "light signal from Mount Palomar arrives on the Moon" is zero.

In the Special Theory of Relativity, the distance between two events does not depend on the reference system. When we change from a stationary system to a moving system, the coordinates of space and time do change—but the distances between events do not. As a consequence, the speed of light does not depend on the system, as Newtonian mechanics would make us suspect: it is a universal quantity. This is so because it constitutes part of the definition of the distance between two points in spacetime. And that is what raises its status to that of a fundamental geometrical quantity.

Given that the relativistic distance does not change when we move from one coordinate system to another one, it is not possible that either the time distance or the space distance will remain unchanged. We can't have it all, as we do in Newton's mechanics, if we want to keep the relativistic distance the same. Experiments tell us quite clearly that the velocity of light remains the same irrespective of the reference system; this means the relativistic distance is invariant—the way it is expected when space and time are treated alike. Space and time are not separate in this context.

But they do change when we make a transition from one system to another one: they are dependent on the observer. That makes us speak of the relativity of space and time. Einstein kept insisting that space and time are not distinct entities: what matters is the unity of space and time that his theory brought to the fore, and which is manifested by the invariance of relativistic distances. That is why he first talked about his theory in terms of a theory of the absolute, not the relative. Nevertheless, the term "relativity theory" did win out in the end. The key feature is the unique relevance of the distance between two spacetime events for the geometric structure of spacetime; and this feature is taken over by the General Theory of Relativity. The phenomenon of gravitation, which finds a successful description in the framework of this theory, is nothing more than a curvature of spacetime, where the relevant geometry is described in terms of the relativistic distance between individual events.

Glossary

absolute zero: zero point of the absolute temperature scale, or $-73.15°$ Celsius. It corresponds to the temperature where matter has no thermal energy.

acceleration: increase of the velocity of an object per unit time.

Anderson, Carl David (1905–1991): American physicist; he discovered the first antiparticle, the positron. Won the Nobel Prize for Physics in 1936. Professor at the California Institute of Technology, Pasadena.

ansatz: mathematical construction, or attempted solution. Trial solution for a mathematical problem (from the German).

antimatter: matter made up of antiparticles.

antiparticle: there is an antiparticle for every particle in Nature. It has the same mass, but all charges have the opposite sign.

antiproton: antiparticle of the proton.

atoms: normal matter is made up of atoms. Atoms consist of a positively charged nucleus, made up in turn of protons and neutrons, and of a shell of negatively charged electrons that orbit around the nucleus.

Becquerel, Antoine Henri (1852–1908): French physicist. He discovered radioactive radiation in 1896 while investigating uranium minerals. Shared the Nobel Prize for Physics in 1903 with Pierre and Marie Curie.

black hole: a region in spacetime where the gravitational pull is so strong that not even light will elude it.

Bolyai, János B. (1802–1860): Hungarian engineer and mathematician. Developed the basics of non-Euclidean geometry independently of Carl Friedrich Gauss and Nikolai Lobachevski.

Bruno, Giordano (1548–1600): Italian natural philosopher. His unified picture of the world was based on a cosmos of infinite extent, contradicting the prevailing dogma of the Catholic Church. He was burned at the stake in Rome.

CERN: acronym for Conseil Européen pour la Recherche Nucléaire, in Geneva. CERN is the largest research laboratory for particle physics in the world. It was founded by the governments of twelve West European countries in 1954.

Chandrasekhar, Subrahmanyan (1910–1995): American astrophysicist of Indian origin. Starting 1942, professor at the University of Chicago. He was the first to study the development of stars using the methods of atomic physics. Co-winner of the 1983 Nobel Prize for Physics, for his work on white dwarfs.

chromodynamics: theory of the forces between quarks, the carrier particles of which are gluons.

cosmological constant: In Einstein's original theory of the universe as a static structure, he introduced the cosmological constant to counteract the gravitational attraction that would eventually have all galaxies collapse. Today, this constant is used as a parameter that helps to describe the global dynamics of the universe.

DESY: acronym for Deutsches Elektronen Synchroton, Hamburg, Germany. It is the German research center for elementary particle physics.

Dirac, Paul Adrien Maurice (1902–1984): British theoretical physicist. Professor at Cambridge University. Initiator of quantum electrodynamics. Shared 1933 Nobel Prize for Physics with Erwin Schrödinger.

Eddington, Arthur Stanley (1882–1944): British astrophysicist and astronomer. Professor at Cambridge University and director of its observatory.

electrodynamics: science describing the electromagnetic phenomena and forces of Nature.

electron: the lightest negatively charged elementary particle. Together with the proton and neutron, one of the building blocks of atoms and thereby of matter. Its mass is 9.109389 times 10^{-28} g, corresponding to 0.511 MeV. It carries one unit of charge.

electron volt: the energy an electron obtains when being accelerated through a voltage difference of 1 volt. *Acronym*: eV, 1 MeV = 1 million eV; 1 GeV = 1 billion eV (from the Greek *gigas*).

elementary particle: in addition to the constituents of atoms, many other particles are known today. Many of them are called "elementary" only for historical reasons, and are now known not to be truly elementary but to consist of quarks and antiquarks instead.

energy: the term used in physics to describe the capability of a system to do work. Energy exists in different forms, such as the kinetic energy of motion. According to the special theory of relativity, energy and mass can be changed into each other. Basic units: 1 joule (J) or watt second (Ws). In particle physics, the energy is usually measured in electron volts (eV).

ether: a hypothetical medium for the transmittal of the forces of Nature that act over a distance, such as gravitation or electromagnetism. Although much quoted in the nineteenth century, it is untenable in the framework of Einstein's relativity theory.

Euclid (c. 450–c. 375 B.C.): Greek philosopher and mathematician, working at the Museion in Alexandria. Author of the *Elements*, the best-known text of Greek mathematics. This book was the principal source for the teaching of mathematics for more than two thousand years.

Euclidean space: a space according to the axioms of Euclid. Specifically, normal three-dimensional space in our daily lives.

event: a point in the spacetime continuum, designated by its coordinates in space and time.

Fermi National Accelerator Laboratory (FNAL; also, Fermilab): U.S. national laboratory for high-energy physics research, located in Batavia, Illinois, near Chicago.

field: an extended physical system pervading space and masses (e.g., field of force, electromagnetic field, solar field, etc.).

Franklin, Benjamin (1706–1790): American politician, natural scientist, and writer. His scientific work was centered on his investigation of electricity (starting 1746), which led to his invention of the lightning rod.

Friedmann, Alexander (1888–1925): Russian mathematician and astrophysicist who, from 1922 to 1924, developed his theory of cosmology on the basis on Einstein's equation of gravitation.

galaxy: a large assembly of stars, held together by gravitational forces.

Galileo (1564–1642): *Full name:* Galileo Galilei; Italian mathematician, physicist, and philosopher. Founder of classical mechanics.

gamma rays: electromagnetic waves of very short wavelengths. They originate in the collision of elementary particles or in radioactive decay.

Gauss, Carl Friedrich (1777–1855): German mathematician, astronomer, and physicist. One of the most versatile mathematicians of all time. His investigations around 1828 on the projection of cards formed the basis of his important theory of planes, differential geometry, and non-Euclidean geometry.

gedanken experiment: a thought experiment; that is, an experiment we imagine we are performing, where our understanding of physics may suggest a solution without the need for actual experimentation (which may be practically either very hard or even impossible).

Gell-Mann, Murray (1929–): U.S. physicist, one of the architects of modern particle physics. He proposed, independently with George Zweig, quarks as the elementary constituents of protons and neutrons. He received the Nobel Prize in 1969.

geodesic line: shortest or longest line that joins two points in a space of a given metric. In Euclidean space, the geodesic lines are straight lines.

gluon: subnuclear particle without mass or electric charge; carrier of the strong nuclear force that binds quarks together.

gravitation: the phenomenon that makes massive objects attract each other. In the general theory of relativity, it is the consequence of the change of space-time structure imposed by the massive object present.

Guericke, Otto von (1602–1686): German natural scientist. Became the mayor of Magdeburg in 1646. One of the pioneers in the research of air pressure and vacuum technology.

Hawking, Stephen (1942–): British astrophysicist. Professor at Cambridge University since 1977. Author of seminal research publications on black holes and cosmology.

Heisenberg, Werner Karl (1901–1976): German theoretical physicist. One of the originators of quantum mechanics. Became professor at Munich University in 1958 and received the Nobel Prize for Physics in 1932.

Higgs particle: hypothetical particle named after the British theoretical physicist Peter Higgs. It carries the field that gives masses to elementary particles.

Hubble, Edwin Powell (1889–1953): American astronomer. Founder of modern extragalactic astronomy. He formulated the principle of cosmic expansion in 1929.

inertial system: a physical reference system where the line of motion of a free object is a straight line.

inflation: hypothetical rapid expansion of the universe that set in shortly after the Big Bang.

Kepler, Johannes (1571–1630): German astronomer and mathematician. Discovered the laws of planetary motion.

K-mesons: unstable mesons that contain an s-quark or s-antiquark. Can be generated in collisions of nucleons or atomic nuclei.

Lemaître, Abbé Georges (1894–1966): Belgian astrophysicist. Canon priest and university professor in Louvain, Belgium.

LEP: acronym for Large Electron Positron accelerator; one of the accelerators at the CERN laboratories in Geneva, Switzerland. It accelerates electrons and positrons to energies of more than 100 GeV before colliding them.

lepton (from the Greek word *leptos* = light): particle with the spin quantum number ´ that does not participate in the strong nuclear interaction. There are three charged leptons: the electron, the muon, and the tau-lepton (Á-lepton). Each has a neutral companion of the same "flavor": the electron-neutrino, the mu-neutrino, and the tau-neutrino (Á-neutrino).

Lichtenberg, Georg Christoph (1742–1799): German physicist and science writer. As of 1770, professor at Göttingen University. He made important contributions to experimental physics, particularly to the theory of electricity and in astronomy.

LHC: acronym for Large Hadron Collider. The largest CERN accelerator, presently under construction. To be installed in the tunnel of the LEP accelerator. It will accelerate protons up to energies of 8,000 GeV before colliding.

Lobachevski, Nicolai Ivanovich (1792–1856): Russian mathematician. As of 1814, professor at Kazan. Along with Bolyai and Gauss, he was one of the discoverers of non-Euclidean geometry.

mass: basic quantity in physics, giving a measure for the inertia of a body that resists changes in its state of motion. *Units:* gram, kilogram.

mesons: strongly interactive particles of integral spin, made up of a quark and an antiquark.

metric: the definition of the measurements in a given space. It fixes the distance between two points in that space.

Minkowski, Hermann (1864–1909): German mathematician. Professor at Zurich and Göttingen universities. He laid the foundations for the mathematical description of relativistic spacetime.

momentum: a quality that characterizes the motion of an object, equal to the product of its mass and its velocity (in the simplest case), introduced by Isaac Newton.

Mössbauer, Rudolf Ludwig (1929–): German physicist. He discovered the effect named after him, where the emission and absorption of energetic photons happens without recoil.

muon: unstable particle related to the electron. Its mass is about 200 times the electron's mass. It decays into an electron, a muon-neutrino, and an electron antineutrino.

neutrino: neutral partner of the charged lepton.

neutron: strongly interacting particle without electric charge. Together with the proton, it is a constituent of atomic nuclei.

neutron star: a cold star consisting of nothing but neutrons.

nuclear force: the force that binds protons and neutrons together in the atomic nucleus. It is now known to be an indirect consequence of the chromodynamic forces between the quarks.

nuclear fusion: the process whereby two atomic nuclei fuse together to form one larger nucleus.

nucleons: collective word for protons and neutrons, the building blocks of the atomic nucleus. Nucleons consist of the positively charged proton and the electrically neutral neutron, and are made up of three deeply bound quarks each.

Oppenheimer, J. Robert (1904–1967): American atomic physicist who was the scientific project leader for the Manhattan Project, the U.S. atomic bomb development project, during World War II, at Los Alamos in New Mexico.

parsec: one parsec (pc) is the distance from which the mean separation of the Earth and the Sun appears under an angle of 1 second of arc ($1/3600$ degrees). 1 pc = 3,087 times 10^{13} kilometers = 3.26 light years.

Pauli, Wolfgang (1900–1958): Swiss-American theoretical physicist, born in Austria. As of 1928, professor in Zurich. Discoverer of the exclusion principle in atomic physics that was named after him. He received the Nobel Prize for Physics in 1945.

Pauli principle: the principle discovered by Wolfgang Pauli, which states that two equal particles of half-integral spin (say, electrons) cannot coexist in the same quantum mechanical state, e.g., they cannot have the same position and the same velocity.

Penrose, Roger (1931–): British mathematician and theoretical physicist. Contributed important work to the general theory of relativity and on black holes.

Penzias, Arno (1933–): U.S. astrophysicist. Codiscoverer, with Robert Wilson, of the cosmic background radiation.

photon: the elementary particle of light. The photon is massless; it carries the electromagnetic force.

Planck, Max (1858–1947): German theoretical physicist. He founded quantum mechanics in 1900, when he defined the laws of radiation that were named after him. As of 1888, a professor in Berlin. He won the Nobel Prize for Physics in 1918.

positron: the antiparticle of the electron; it carries one positive elementary charge.

proton: the positively charged elementary particle that is also the nucleus of the hydrogen atom. All other nuclei consist of protons plus neutrons.

quantum mechanics: the theory of quantum phenomena in mechanics. It is at the basis of modern atomic physics.

quantum theory: the theory of quantum phenomena, particular in atomic physics.

quarks: building blocks of the nucleon: protons and neutrons are made up of three quarks each. Six quark types—up and down, charm and strange, top and bottom—have been identified, of which only two, the u-quark and d-quarks, act as constituents of stable matter. The other four types (or, in the jargon of the field, "flavors") make up new unstable particles. Quarks have mass, which is very small for u-quarks and d-quarks. The heaviest quark, the t-quark, has a mass of about 180 GeV, making it the heaviest known elementary object to date. Quarks were introduced as a concept in 1964 by, independently, Murray Gell-Mann and George Zweig (Zweig called them aces, not quarks); they were to explain the observed symmetries of elementary particles in a simple way.

quasar: very luminous center of a very distant galaxy.

radioactivity: spontaneous change of a number of atomic nuclei which, while emitting energetic particles or rays, change into other nuclei, setting energy free in the process. They emit alpha rays (helium nuclei, consisting of two neutrons and two protons), beta rays (consisting of electrons or positrons), or gamma rays (photons).

Riemann, Bernhard (1826–1866): German mathematician, professor at Göttingen University starting 1859. In the classic paper that earned him his professorship, "Über die Hypothesen, welche der Geometric zugrunde liegen" (The hypotheses that are the foundations of geometry), published posthumously in 1867, he laid down the rules Einstein used later on to shape his general theory of relativity.

Russell, Bertrand (1872–1970): British mathematician, philosopher, and writer. He received the Nobel Prize for Literature in 1950.

Sandage, Allan (1926–): American astronomer, author of seminal work on extragalactic astronomy.

Schwarzschild, Karl (1873–1916): German astronomer, professor of astronomy at Göttingen University as of 1902. Starting 1909, director of the Astrophysical Observatory in Potsdam outside Berlin. Author of important treatises on photometry and on the orbits of stellar objects. In 1915 he discovered the solutions to Einstein's equations on general relativity which are named after him.

second of arc: angular unit ($1/3600$ part of a degree, $1/60$ part of a minute of arc).

Sommerfeld, Arnold (1868–1953): German theoretical physicist. Author of influential papers on atomic physics. Starting 1906, professor at Munich University.

supernova: exploding star which ejects most of its stellar matter into the universe.

tensor: a generalization of the mathematical term *vector* (see below). A vector is a tensor of the first rank. A tensor of the second rank is formed by the multiplication of the components of a vector.

uncertainty principle: a rule discovered by Werner Heisenberg which says that it is impossible to denote both the position and the momentum (velocity) of a particle with arbitrary precision.

vector: a quantity that is defined both by magnitude and by direction. In three-dimensional space, the vector is indicated by an arrow, as defined by its three components.

W particle: fundamental particle which, together with the Z particle (see below), is the carrier of the weak interaction. There are both positively and negatively charged W bosons; they form a particle-antiparticle pair. Their mass is about 81 GeV.

Wheeler, John Archibald (1911–): American nuclear physicist, author of influential papers on astrophysics and black holes. During and after World War II, he was an influential contributor to the development of the hydrogen bomb.

white dwarf: a cold star, the stability of which is guaranteed by the Pauli principle acting between the electrons.

Wigner, Eugene Paul (1902–1994): American theoretical physicist, professor at Princeton University starting in 1938. Author of an important work on the principles of symmetry.

Wilson, Robert W. (1936–): American astrophysicist; codiscoverer, with Arno Penzias, of the cosmic background radiation.

world-line: in spacetime, a moving object defines a line—the sequence of events through which the object passes. One such line is known as a world-line; it contains all the information about motion of an object in the past, the present, and the future.

Z particle: elementary particle which, together with the W particle (see above), mediates the weak interaction. Electrically neutral, it has a mass of about 91 GeV.

Zeldovich, Yakov: (1914–1987): Russian astrophysicist, author of important work on black holes. Top contributor to the Russian hydrogen bomb project.

Zwicky, Fritz (1898–1974): Swiss-born U.S. astrophysicist and astronomer. Starting 1927, professor at the California Institute of Technology in Pasadena. He predicted the existence of neutron stars.

Notes

1. Introduction

1. Albert Einstein, quoted in Albrecht Fölsing, *Albert Einstein: A Biography* (Frankfurt am Main, 1993), 417 [*Trans.*: my translation from the German ed.]; English version, New York: Penguin, 1998), n.p.
2. Einstein, in his article on the foundations of the general theory of relativity, in *Annals of Physics* 51, 4th ser. (1916): 639–42 [*Trans.*: my translation]; also cited in Fölsing, *Albert Einstein* (Frankfurt am Main, 1993), 535.
3. Einstein to James Franck, quoted in C. Seelig, *Albert Einstein* (Zurich, 1960), 72 [*Trans:* my translation from the German].

2. Meeting Einstein and Newton in Caputh

1. Quoted in B. Hoffman, *Albert Einstein* (Zurich, 1976), 164 [*Trans.*: my translation from the German].

3. "Subtle is the Lord"

1. Quoted in B. Hoffman, *Albert Einstein* (Zurich, 1976), 172 [*Trans.*: my translation from the German].

4. Particles and Their Masses

1. Quoted in C. Seelig, *Albert Einstein* (Zurich, 1960), 397 [*Trans.*: my translation].

5. Haller's Lecture: Empty Space and Modern Physics

1. Quoted in B. Hoffman, *Albert Einstein* (Zurich, 1976), 297 [*Trans.*: my translation from the German].

2. This chapter is a slightly reworked version of an article that originally appeared in the German *P.-M. Magazine* (November 1991): 40–54, and is used here with the kind permission of *P.-M. Magazine*.

6. What Is Mass?

1. Quoted in C. Seelig, *Albert Einstein* (Zurich, 1960), 281 [*Trans.*: my translation from the German].

7. Gravity—Is It a Force?

1. From Albert Einstein, "The Origins of the General Theory of Relativity" (Gibson Lecture at the University of Glasgow, June 20, 1933; University of Glasgow Publication No. 20), cited in Fölsing, *Albert Einstein* (Frankfurt am Main, 1993), 419 [*Trans.*: my translation from the German].

8. Does Light Bend?

1. Albert Einstein, *Mein Weltbild* (Zurich, 1953), 173 [*Trans.*: my translation from the German]; English version, *The World As I See It* (reissue; Sacramento: Citadel, 1993).
2. Thomson, quoted in Fölsing, *Albert Einstein* (Frankfurt am Main, 1993), 501 [*Trans.*: my translation from the German].

9. A Flat World Curved

1. Quoted in C. Seelig, ed., *Helle Zeit, dunkel Zeit* (Zurich, 1956), 73 [*Trans.*: my translation from the German].

10. Curved Space and Cosmic Laziness

1. Quoted in C. Seelig, *Albert Einstein* (Zurich, 1960), 119 [*Trans.*: my translation from the German].

11. Time Bent

1. Einstein, letter to Walter Dällenbach dated March 31, 1915; cited in Fölsing, *Albert Einstein* (Frankfurt am Main, 1993), 414 [*Trans.*: my translation from the German].
2. Rudolf Keyser, quoted in M. Grüning, *Ein Haus für Albert Einstein* [A house for Albert Einstein] (Berlin, 1990), 469 [*Trans.*: my translation from the German].
3. Wigner, cited in ibid., 528 [*Trans.*: my translation from the German].

12. Matter in Space and Time

1. Einstein, quoted in A. Sommerfeld, *Gesammelte Schriften* 4 (1968): 646 [*Trans.*: my translation from the German].

13. A Star Bends Space and Time

1. Quoted in C. Seelig, *Albert Einstein* (Zurich, 1960), 304 [*Trans.*: my translation from the German].

14. The Cemetery of Stars

1. Albert Einstein, *Briefe* [Correspondence] (Zurich, 1981), 34 [*Trans.*: my translation from the German].

15. The Wall of Frozen Time

1. A. Einstein and M. Besso, *Correspondence, 1903–1955* (Paris, 1972), 538 [*Trans.*: my translation from the German].

16. In the Atrium of Hell

1. Quoted in B. Hoffman, *Albert Einstein* (Zurich, 1976), 17 [*Trans.*: my translation from the German].

17. The Monster of Spacetime

1. Einstein, quoted in P. A. von Schilpp, ed., *Albert Einstein als Philosoph und Naturforscher* (Stuttgart, 1955), 155 [*Trans.*: My translation from the German].

18. The Cosmic Beat

1. Albert Einstein, *Briefe* [Correspondence] (Zurich, 1981), 24 [*Trans.*: my translation from the German]; English version, *Out of My Later Years* (paperback reprint; Westport, Conn.: Greenwood, 1995).

19. The Dynamic Universe and a Blunder of Einstein's

1. Einstein, quoted in P. C. von Aichelburg and R. U. Sexl, eds., *Albert Einstein* (Brunswick [Braunschweig], 1979), 57 [*Trans.*: my translation from the German].

1. Albert Einstein, *Aus meinen späteren Jahren* (Stuttgart, 1979), 56 [*Trans.*: my translation from the German].

21. Echo of the Big Bang

1. Quoted in C. Seelig, *Albert Einstein* (Zurich, 1960), 318 [*Trans.*: my translation from the German].

22. The First Seconds

1. Quoted in C. Seelig, ed., *Helle Zeit, dunkel Zeit* (Zurich, 1956), ch. 22 [*Trans.*: my translation from the German].

23. A Cosmic Fairytale

1. Quoted in C. Seelig, *Albert Einstein* (Zurich, 1960), 336 [*Trans.*: my translation from the German].

SELECTED BIBLIOGRAPHY

BOOKS ABOUT THE RELATIVITY OF SPACE AND TIME

Fritzsch, Harald. *An Equation That Changed the World: Newton, Einstein, and the Theory of Relativity*. Chicago: University of Chicago Press, 1997.

Greene, Brian. *The Elegant Universe: Superstrings, Hidden Dimensions, and the Quest for the Ultimate Theory*. New York: Vintage, 2000.

Hawking, Stephen. *A Brief History of Time*. New York: Bantam, 1998.

Hey, Tony, Patrick Walters, and Anthony J. G. Hey. *Einstein's Mirror*. New York and Cambridge: Cambridge University Press, 1997.

Kaku, Michio. *Hyperspace: A Scientific Odyssey Through Parallel Universes, Time Warps, and the Tenth Dimension*. New York: Anchor Doubleday, 1995.

Pais, Abraham. *Subtle Is the Lord: The Science and the Life of Albert Einstein*. New York and Oxford: Oxford University Press, 1983.

Thorne, Kip. *Black Holes and Time Warps: Einstein's Outrageous Legacy*. New York: Norton, 1997.

BOOKS ABOUT COSMOLOGY

Barrow, John. *The Origin of the Universe*. New York: HarperCollins, 1997.

Ferris, Timothy. *The Whole Shebang: A State-of-the-Universe(s) Report*. New York: Simon and Schuster, 1997.

Fritzsch, Harald. *The Creation of Matter: The Universe from Beginning to End*. New York: Basic Books, 1986.

Guth, Alan H. and Alan P. Lightman. *The Inflationary Universe: The Quest for a New Theory of Cosmic Origins*. Cambridge, Mass.: Perseus, 2000.

Harrison, Edward. *Cosmology: The Science of the Universe*. New York and Cambridge: Cambridge University Press, 2000.

Kolb, Edward W. *Blind Watchers of the Sky: The People and Ideas That Shaped Our View of the Universe*. Cambridge, Mass.: Perseus, 1997.

Overbye, Dennis. *Lonely Hearts of the Cosmos: The Scientific Quest for the Secret of the Universe.* Boston: Little, Brown, 1999.

Rees, Martin. *Before the Beginning: Our Universe and Others.* Reading, Mass.: Addison-Wesley, 1997.

Weinberg, Steven. *The First Three Minutes: A Modern View of the Origin of the Universe.* 2d ed. New York: Basic Books, 1993.

Index

Page numbers in **boldface** type refer to illustrations.

photons: in annihilation of matter and antimatter, 34, 280; defined, 317; electric charge and, 74, 282; electrical forces and, 38; electromagnetic forces and, 270; energy of, 252; as gamma rays, 318; versus gluons, 79; mass of, 34, 37, 54, 79; particle decay and, 292; radiation, 54, 56, 70, 218–19, 263, 264, 266, 274, 282; ratio to nucleons, 266, 274; in vacuum, 54, 64; wavelengths of, **55**

pi-mesons, 45

Planck, Max: background of, 317–18; quantum theory ideas, 1, 203–204, 263

Planck Length (Planck's elementary unit of length), 204, 221, 263, **265**, 293, 303

Planck Time, 204, 303

plane wave, 229–30, **230**

planetary perihelions, 4–5, 7–8, 72–73

positrons, 286; annihilation and, 34–35, **35**, 40–41, **41**, 46, 62–64, **63**, **66**; as beta rays, 318; defined, 318; discovery of, 63, 313; mass of, 64; symmetry and, 291; in vacuum, 61–66, **63**. *See also* LEP (Large Electron Positron) accelerator

primary black holes, 222

Principia. See Philosophiae Naturalis Principia Mathematica (Newton)

protons: acceleration with LHC, 75–76; antiparticles in, 27 (*see also* antiprotons); proton-antiproton annihilation, 48, **50**, 280, 285, 308; Big Bang and numbers of, 266, 280; decay, 181–82, 288, **292**, 292–93, 295, 299; defined, 318; electric charge of, 30, 31; electrical forces and, 25; gluons and, 78, 79–80, **80**; gravitational fields, 37; mass of, 37, 75–80; neutron conversion and, 181–82, **183**, 266–67; neutron decay and, 180, **181**;

quarks and, **80**, 80–81, 268, 269; stability of, 288, 292

Prussian Academy of Sciences, 5, 157, 164

pulsars, 186

Pythagoras's theorem, 115, **115**, 120

quantum field theory, 64–66

quantum mechanics: black holes and, 218–19; defined, 318; gravitation and effects of, 202, 203

quantum theory: defined, 1, 318; gravitation theory and, 204–206; vacuum and, 57–60, 61, 63, 70

quarks: after Big Bang, 280, 288; antiparticles of, 27, **66**, 280, 288; atomic nuclei and, 48, 181, 182; color and, 290; decay of, 270; defined, 318; discovery of, 48, **49**; energy of, 80–81; gas, 182, 183; gluons and, 80–81, 268, 270, 280; gravitational collapse and, 188; mass of, 79, 81, 269–70; matter as, 205, 269; nuclear forces and, 78, 79–80, **80**, 268, 317; nucleons and, 268, 269, 281; production of, **66**; protons and, **80**, 80–81, 268, 269; types of, 268–70, 318. *See also specific quarks*

quasars: black holes in, 215; defined, 318; deflection of light images and, 175, **176**; energy of, 213–14, **214**; wave emissions of, 213

radiation: background, 263 (*see* background radiation); cold, 263; of electric charge, 224; electromagnetic, 80–81 (*see* electromagnetic radiation); electron-positron annihilation and, 34–35; of gravitational waves, 225–26, **228**; isotropic, 264, 266; photon, 54, 56, 70, 218–19, 263, 264, 266, 274, 282; electromagnetic, 173, 213; thermal, 262–63, 299

176; energy density and, 158–59; geodesic lines in, 173 (*see also* geodesic lines); inertial systems and, 88–89; mass and, 223; planar, 149–50, 201; twin paradox and, 2; world-lines in (*see* world-lines)

special theory of relativity: basic tenets of, 307–11; consequences of, 2, 7–8; distance equations in, 147, 310; ether and, 92; mass in, 2; Newtonian mechanics and consequences of, 2, 7–8, 307–8; Newtonian mechanics commonalities with, 3; reference systems in, 308; space in, 22; terminology of, 97; time in, 22, 128–31, 308; velocities in, 231

speed of light, 29, 46–47; constant of, 140; distance between events and, 310–11; escape velocity as, 194; gravitation waves and, 223–24; photons and, 54; velocity and, 29, 46–47, 225, 274, 308. *See also* light

spirals: as geodesic lines, 154

stability of universe: assumptions about, 243–44

Standard Model of Gravitational Theory, 276–77

Standard Model of Particle Physics, 276–77, **277**, 279, 288

stars: black holes and two-star systems, 208–209, 225; density of, 180; gravitation and, 184; gravitational collapse and, 188, 189–98, **200**, 205; time warp during gravitational collapse of, 191–93. *See also* neutron stars

Steinberger, Jack, **43**

strange quarks. *See* s-quarks

Sun: death of, 184–85; dynamical equilibrium of, 184–85; effects of gravitation of, 7–8, 47, 174, 184; electromagnetic waves deflection by, 173; as fusion reactor, 184; mass of, 162, 169

supernovae: defined, 319; energy in, 2, 226; predicting explosions of, 231

symmetry: broken, 71, 300, **300**; of matter, 291, **291**, 293; between matter and antimatter, 64–65, 69, 286, 287, 295; spherical, **166**, 189–90, 201–202

"Systeme du Monde, Le" (The World System) (Laplace), 194–95

Tagore, Rabindranath, 14

tangents, 113–14, **114**, 127, **128**, 146

tau-leptons (t-leptons), 56, 271, 316

tau-neutrinos (t-neutrinos), 56, 275, 316

tauons, 271

temperature: after Big Bang, 279, 302; of black holes, 220 21; of early universe, 262–63; of neutrino gas, 282; of photon radiation, 264, 266, 274, 282

tensor of curvature: defined, 120–21; matter tensor and, 158, 159, 202; metric tensor and, 125, 197

tensors: defined, 319; matter, 158–59, 161, 202–203; metric, 119–20, 125, 127, 147–48, 157–58, 197

TEVATRON accelerator, 70, 308

Thomson, Joseph John, 105

thought experiments. *See* gedanken experiments

time: absolute, 22, 307; in coordinate systems, 128–29, **129**, 309; curvature of, 5, 170; dilation of, 45, 130–31, 138–43, 144, 147–48, **152**, 152–53; equation for dilation of, 147–48; events and proper, 133–34, **134**, 171; events connected by world-lines and proper, 133–34, **134**, 171; "freezing" of, 171, 193; gravitation and flow of, 138–39, 145, 148–49, **171**; gravitation and warping of, 139–40, **141**, 142–43, 170–76, 191–93; Harvard experi-